*Editors*

F.W. Gehring
P.R. Halmos

## Universitext

Editors: F.W. Gehring, P.R. Halmos

**Booss/Bleecker:** Topology and Analysis
**Charlap:** Bieberbach Groups and Flat Manifolds
**Chern:** Complex Manifolds Without Potential Theory
**Chorin/Marsden:** A Mathematical Introduction to Fluid Mechanics
**Cohn:** A Classical Invitation to Algebraic Numbers and Class Fields
**Curtis:** Matrix Groups, 2nd. ed.
**van Dalen:** Logic and Structure
**Devlin:** Fundamentals of Contemporary Set Theory
**Edwards:** A Formal Background to Mathematics I a/b
**Edwards:** A Formal Background to Higher Mathematics II a/b
**Endler:** Valuation Theory
**Frauenthal:** Mathematical Modeling in Epidemiology
**Gardiner:** A First Course in Group Theory
**Godbillon:** Dynamical Systems on Surfaces
**Greub:** Multilinear Algebra
**Hermes:** Introduction to Mathematical Logic
**Hurwitz/Kritikos:** Lectures on Number Theory
**Kelly/Matthews:** The Non-Euclidean, The Hyperbolic Plane
**Kostrikin:** Introduction to Algebra
**Luecking/Rubel:** Complex Analysis: A Functional Analysis Approach
**Lu:** Singularity Theory and an Introduction to Catastrophe Theory
**Marcus:** Number Fields
**McCarthy:** Introduction to Arithmetical Functions
**Meyer:** Essential Mathematics for Applied Fields
**Moise:** Introductory Problem Course in Analysis and Topology
**Øksendal:** Stochastic Differential Equations
**Porter/Woods:** Extensions of Hausdorff Spaces
**Rees:** Notes on Geometry
**Reisel:** Elementary Theory of Metric Spaces
**Rey:** Introduction to Robust and Quasi-Robust Statistical Methods
**Rickart:** Natural Function Algebras
**Schreiber:** Differential Forms
**Smoryński:** Self-Reference and Modal Logic
**Stanisić:** The Mathematical Theory of Turbulence
**Stroock:** An Introduction to the Theory of Large Deviations
**Sunder:** An Invitation to von Neumann Algebras
**Tolle:** Optimization Methods

V.S. Sunder

# An Invitation to von Neumann Algebras

Springer-Verlag
New York Berlin Heidelberg
London Paris Tokyo

V. S. Sunder
Indian Statistical Institute
New Delhi-110016
India

AMS Classification: 46-01

Library of Congress Cataloging in Publication Data
Sunder, V. S.
An invitation to von Neumann algebras.
(Universitext)
Bibliography: p.
Includes index.
1. von Neumann algebras. I. Title.
QA326.S86 1986 512'.55 86-10058



Printed and bound by R.R. Donnelley and Sons, Harrisonburg, Virginia.
Printed in the United States of America.

9 8 7 6 5 4 3 2 1

ISBN 0-387-96356-1 Springer-Verlag New York Berlin Heidelberg
ISBN 3-540-96356-1 Springer-Verlag Berlin Heidelberg New York

# PREFACE

Why This Book:

The theory of von Neumann algebras has been growing in leaps and bounds in the last 20 years. It has always had strong connections with ergodic theory and mathematical physics. It is now beginning to make contact with other areas such as differential geometry and $K$-Theory. There seems to be a strong case for putting together a book which (a) introduces a reader to some of the basic theory needed to appreciate the recent advances, without getting bogged down by too much technical detail; (b) makes minimal assumptions on the reader's background; and (c) is small enough in size to not test the stamina and patience of the reader. This book tries to meet these requirements. In any case, it is just what its title proclaims it to be -- an invitation to the exciting world of von Neumann algebras. It is hoped that after perusing this book, the reader might be tempted to fill in the numerous (and technically, capacious) gaps in this exposition, and to delve further into the depths of the theory.

For the expert, it suffices to mention here that after some preliminaries, the book commences with the Murray - von Neumann classification of factors, proceeds through the basic modular theory to the $III_\lambda$ classification of Connes, and concludes with a discussion of crossed-products, Krieger's ratio set, examples of factors, and Takesaki's duality theorem. Although the material is standard, some of the treatment (particularly in Sections 4.1 - 4.3) may be new.

Shortcuts taken:

In order to accommodate all the above-mentioned material in a volume this size, it was necessary to take some shortcuts:

(i) Some theorems, though stated in full generality, are only proved under additional (sometimes very severe) simplifying assumptions -- typically, to the effect that some operator is bounded. Some other results suffer a sorrier fate -- they are not even graced with an apology for a proof.
(ii) Arguments of a purely set-topological nature often receive step-motherly treatment; where the argument is painless, it has been included; where it is not, the reader is entreated to accept, in good faith, the validity of the relevant statement.
(iii) The exercises are an integral part of the book. Several "lemmas" have been relegated to the exercises; any exercise, which is even slightly non-obvious, is furnished with "hints", which are often more in the nature of outlines of solutions. The exercises, rather than being compiled at ends of sections, punctuate the text at junctures where they seem to fit in most naturally.
(iv) Both exercises and unproved results are treated just like properly established theorems, in that they are unabashedly used in subsequent portions of the text.

The prospective reader:

This book is aimed at two classes of readers: graduate students with a reasonably firm background in analysis, as well as mature mathematicians working in other areas of mathematics. As a matter of fact, this book grew out of a course of (twelve) lectures given by the author while visiting the Indian Statistical Institute at Calcutta in the summer of 1984. It was largely due to the positive response of that audience -- consisting entirely of members of the second category mentioned above -- that the author embarked on this venture.

The reader is assumed to be familiar with elementary aspects of:

(a) measure theory -- monotone convergence, Fubini's Theorem, absolute continuity, $L^p$ spaces for $p = 1,2,\infty$;
(b) analytic functions of one complex variable -- sparseness of zero-sets, contour integration, theorems of Cauchy, Morera, and Liouville;
(c) functional analysis -- the "three principles", weak and weak* topologies;
(d) Hilbert spaces and operators -- orthonormal basis, subspaces and projections, bounded operators, self-adjoint operators. (The necessary background material from Hilbert space theory is rapidly surveyed in Section 0.1.)

In the latter part of the book, a nodding acquaintance with abstract harmonic analysis will be helpful, although it is not essential. For the reader who has been denied such a pleasure, a brief appendix (on topological groups) should serve to perform the necessary introduction, which should precede the furtherance of that acquaintance in Sections 3.2 and 3.3. An attempt has also been made, in Section 3.2, to compile the necessary results from the theory before proceeding to use them.

Trappings:

This volume is equipped with some of the standard fittings, such as a list of symbols, an index of terms used, some notes of a bibliographical nature, and a bibliography. The bibliographical notes are somewhat terse; for more details, the reader may consult [Tak 4]. The terseness also extends to the bibliography, which lists only those books and papers that bear directly on the treatment here; for an extensive bibliography, the reader might consult [Dix]. If the reader spots some inaccuracy in the notes or the references, or anywhere else in the text for that matter, the author would appreciate being informed of such an error.

The title:

The author would like to take this opportunity to thank Professor Arveson for kindly permitting the use of a title that is highly reminiscent of his delightful little book on $C^*$-algebras. If this volume manages to capture even a miniscule fraction of the charm displayed in that volume, it would have accomplished all that the author could have hoped for.

# ACKNOWLEDGMENTS

I would like to thank the following people for the roles they have played in the production of this book: Professor A. K. Roy, for having invited me to spend six wonderful weeks at Calcutta; the entire audience for the course of lectures I gave at Calcutta, for their enthusiasm and positive response; Professor M. G. Nadkarni, for some discussions concerning Krieger's ratio set; Krishna, for having faithfully and enthusiastically attended all those seminars I organized, whereby I learnt the theory of von Neumann algebras; Shobha Madan, for painstakingly reading large portions of the manuscript and picking out several errors; Professor W. Arveson for a very encouraging letter which boosted my sagging morale at a crucial stage; Shri V. P. Sharma, for an extremely efficient job of typing, cheerfully performed in an amazingly short period of time; and finally, Vyjayanthi, for reasons too uncountable to enumerate, and to whom this book is fondly dedicated.

# CONTENTS

# LIST OF SYMBOLS

The accompanying number usually refers to the page on which the symbol first occurs, or is explained. The symbols are arranged in the order of their appearance in the text.

# Chapter 0
# INTRODUCTION

As the title suggests, this chapter is devoted to developing some of the basic technical results needed in the theory, and may be safely omitted by the expert. The first section establishes some of the notation employed throughout the book and lists, without proof, the basic facts concerning operators on Hilbert space. The next section establishes the "non-commutative analogue" of the classical results $c_0^* = \ell^1$ and $(\ell^1)^* = \ell^\infty$ -- null-convergent, summable, and bounded sequences being replaced by compact, trace-class and bounded operators respectively.

The existence of a predual yields a locally convex topology -- namely, the weak* topology -- with respect to which the unit ball of $\mathfrak{L}(\mathcal{H})$ is compact. It would not be out of place to suggest that this is one of the primary reasons for the richness of the theory of von Neumann algebras. The third section is a brief examination of this ($\sigma$-weak) topology, as well as of the more easily defined strong and weak topologies. Some other topologies -- such as the strong* and $\sigma$-strong* topologies -- are relegated to the exercises.

The final section contains the fundamental double-commutant theorem of von Neumann, the definition, and some elementary properties of von Neumann algebras.

## 0.1. Basic Operator Theory

A von Neumann algebra, the central topic of this book, is a certain kind of a set of operators on a Hilbert space. Throughout the book, only complex separable Hilbert spaces are considered; the adjectives "complex" and "separable" will almost never be stated, but it will be tacitly assumed that they apply. Hilbert spaces will be usually denoted by $\mathcal{H}$; on a few occasions, where more than one Hilbert space is under consideration, the symbols $\mathcal{K}$ and $\mathcal{M}$ are also employed. Further, we shall adhere to the mathematical (rather than the physical) convention whereby inner products are linear in the first variable and conjugate - linear in the second (rather than the other way around).

Consistent with our resolution to totally disregard nonseparable Hilbert spaces, we shall only consider measure spaces if they are separable. Actually, we shall only consider measure spaces $(X,\mathcal{F},\mu)$ where $\mu$ is a non-negative $\sigma$-finite measure space, such that $L^2(X,\mu)$ is separable.

Subspaces of $\mathcal{H}$ will usually be denoted by symbols such as $\mathcal{M}$ and $\mathcal{N}$. If $\{\mathcal{M}_n\}_{n=1}^{\infty}$ is a sequence of closed subspaces of $\mathcal{H}$, and if $\mathcal{M}_n \perp \mathcal{M}_m$ for $n \neq m$, we shall write $\oplus_{n=1}^{\infty}\mathcal{M}_n$ for the closure of $\Sigma_{n=1}^{\infty}\mathcal{M}_n$; the "direct sum" notation will be used only for an orthogonal direct sum of closed subspaces. Of course, we shall also write $\oplus_{n=1}^{\infty}\mathcal{H}_n$ for the "external" direct sum of Hilbert spaces, in which case each $\mathcal{H}_n$ will be naturally identified with a subspace of the direct sum. If $\mathcal{M}$ and $\mathcal{N}$ are closed subspaces of $\mathcal{H}$ with $\mathcal{N} \subseteq \mathcal{M}$, we shall write $\mathcal{M} \ominus \mathcal{N}$ for $\mathcal{M} \cap \mathcal{N}^{\perp}$.

Vectors in $\mathcal{H}$ will be denoted by $\xi,\eta,\zeta$, etc., while symbols such as $a,x,y,z,e,f,u,w$ will always denote bounded operators. It will be necessary, on occasion, to consider unbounded operators, such objects being usually denoted by $A$, $H$, $K$, $S$, $F$, etc. Of course, it may turn out in some instances that $S$ is actually bounded; when that happens, the consequent relief would, it is hoped, offset the conflict with our notational convention.

The set $\mathcal{L}(\mathcal{H})$ of all bounded linear operators on $\mathcal{H}$ has the structure of a $C^*$-algebra: explicitly, it is a Banach algebra (with respect to the operator norm $\|x\| = \sup\{\|x\xi\|: \xi \in \mathcal{H}, \|\xi\| = 1\}$, pointwise vector operations and composition product), equipped with an involution $x \to x^*$, which satisfies the so-called $C^*$-identity: $\|x^*x\| = \|x\|^2$.

The orthogonal projection associated with a closed subspace $\mathcal{M}$ will usually be denoted by $p_{\mathcal{M}}$; this is the operator satisfying $p_{\mathcal{M}} = p_{\mathcal{M}}^2 = p_{\mathcal{M}}^*$ and ran $p_{\mathcal{M}} = \mathcal{M}$. (Here and in the sequel, the range of an operator $x$ will be denoted by ran $x$.) Conversely any operator $p$ satisfying $p = p^2 = p^*$ is the orthogonal projection onto ran $p$. Such operators will be simply referred to as projections. We shall never consider non-self-adjoint projections. Recall that the operator $x$ is called self-adjoint if $x = x^*$; more generally, for any set $M \subseteq \mathcal{L}(\mathcal{H})$, we shall let $M^* = \{x^*: x \in M\}$ and call $M$ self-adjoint if $M = M^*$.

Probably the most fundamental theorem in Hilbert space theory is the spectral theorem for self-adjoint operators, which may be formulated thus: let $x$ be a self-adjoint operator with spectrum sp $x$; then there exists a mapping $F \to e(F)$ from the class of Borel subsets of sp $x$ to the class of projections in $\mathcal{H}$ satisfying: (a) $e(\text{sp}\, x) = 1$; (b) if $F = \cup_{n=1}^{\infty}F_n$ and $F_n \cap F_m = \phi$ for $n \neq m$, then $\{e(F_n)\}_{n=1}^{\infty}$ is a sequence of pairwise orthogonal projections and ran $e(F) = \oplus_{n=1}^{\infty}$ran $e(F_n)$; and (c) for any $\xi,\eta$ in $\mathcal{H}$, if $\mu_{\xi,\eta}$ is the finite complex measure on sp $x$ defined by $\mu_{\xi,\eta}(F) = \langle e(F)\xi,\eta\rangle$, then $\langle x\xi,\eta\rangle = \int \lambda d\mu_{\xi,\eta}(\lambda)$.

The preceding two statements are usually paraphrased thus: if $x = x^*$, there exists a spectral measure $e(\cdot)$ defined on sp $x$ such that $x = \int \lambda \, de(\lambda)$.

In fact, more is true: let $\mu$ be any measure on sp $x$ such that $\mu(F) = 0$ iff $e(F) = 0$. (One such measure is given by

$$\mu(F) = \sum_n 2^{-n} \langle e(F)\xi_n, \xi_n \rangle,$$

where $\{\xi_n\}$ is an orthonormal basis for $\mathcal{H}$.) Then there exists an isometric algebra isomorphism $f \to f(x)$ from $L^\infty(\text{sp } x, \mu)$ into $\mathcal{L}(\mathcal{H})$ such that for any $\xi, \eta$ in $\mathcal{H}$, $\langle f(x)\xi, \eta \rangle = \int f(\lambda) d\mu_{\xi,\eta}(\lambda)$, with $\mu_{\xi,\eta}$ as in (c) above; further, $f(x)^* = \bar{f}(x)$. It should be clear that $e(F) = 1_F(x)$, where $1_F$ denotes (here and throughout the book) the indicator or characteristic function of $F$.

An operator $x$ is said to be positive (denoted $x \geqslant 0$) if $\langle x\xi, \xi \rangle \geqslant 0$ for all $\xi$ in $\mathcal{H}$, or, equivalently, if $x = x^*$ and sp $x \subseteq [0,\infty)$. (The equivalence of these conditions is proved using the spectral theorem and the polarization identity which asserts that if $x \in \mathcal{L}(\mathcal{H})$ and $\xi, \eta \in \mathcal{H}$, then

$$4\langle x\xi, \eta \rangle = \sum_{k=0}^{3} i^k \langle x(\xi + i^k\eta), \xi + i^k\eta \rangle.)$$

If $x \geqslant 0$, there exists a unique $y \geqslant 0$ such that $x = y^2$; the square root $y$ is given by $y = f(x)$, where $f(t) = t^{1/2}$ for $t \geqslant 0$, and will always be denoted by $y = x^{1/2}$.

The spectral theorem, as we have stated, is valid for the larger class of normal operators, an operator $x$ being called normal if $x^*x = xx^*$. An important class of normal operators is the class of unitary operators. Recall that an operator $u$ is unitary iff $u^*u = uu^* = 1$. (Here and elsewhere, the identity operator is denoted simply by 1, and $\lambda \cdot 1$ is noted simply by $\lambda$.)

An operator $u$ satisfying $u^*u = 1$ (resp., $uu^* = 1$) is called an isometry (resp., a coisometry). More generally, an operator $u$ is called a partial isometry if $\|u\xi\| = \|\xi\|$ whenever $\xi \in \ker^\perp u$. It is well-known that $u$ is a partial isometry if and only if $e = u^*u$ is a projection; in this case, $f = uu^*$ is also a projection and the subspaces ran $e$ (= ran $u^*$) and ran $f$ (= ran $u$) are called, respectively, the initial and final spaces of the partial isometry $u$.

Another basic result is the polar decomposition theorem, which states that every operator $x$ admits a decomposition $x = uh$ which is uniquely determined by the following conditions: $u$ is a partial isometry, $h \geqslant 0$ and ker $u$ = ker $h$ = ker $x$. The positive factor $h$ is given by $h = (x^* x)^{1/2}$ and will usually be denoted by $|x|$. It is not hard to show that if $x = u|x|$ is the polar decomposition of $x$, then $x^* = u^*|x^*|$ is the polar decomposition of $x^*$; more generally, $(uf(|x|))^* = u^* \bar{f}(|x^*|)$ for any Borel function $f$ on $[0,\infty)$.

So much for bounded operators; let us now recall some facts concerning unbounded operators. A (possibly unbounded) operator is a linear map $H: \mathfrak{D} \to \mathcal{H}$ where $\mathfrak{D}$ is some linear (not necessarily closed) subspace of $\mathcal{H}$, called the domain of $H$ and denoted by dom $H$. The operator $H$ is called a closed operator if it has a closed graph, i.e., if $G(H) = \{(\xi,H\xi): \xi \in \text{dom } H\}$ is a closed subspace of $\mathcal{H} \oplus \mathcal{H}$.

For a densely defined operator $H$ (i.e., dom $H = \mathcal{H}$), the adjoint $H^*$ is the uniquely defined linear operator with domain given by dom $H^* = \{\eta \in \mathcal{H}: \exists\, c > 0 \ni |\langle H\xi,\eta\rangle| \leqslant c\|\xi\| \ \forall \xi \in \text{dom } H\}$ and satisfying $\langle H\xi,\eta\rangle = \langle \xi,H^*\eta\rangle$ whenever $\xi \in \text{dom } H$ and $\eta \in \text{dom } H^*$.

An operator $T$ is called an extension of an operator $S$ if $G(S) \subseteq G(T)$, i.e, if dom $S \subseteq$ dom $T$ and $T\xi = S\xi$ for $\xi$ in dom $S$. The equation $S = T$ is to be interpreted as $S \subseteq T$ and $T \subseteq S$.

An operator $S$ is said to be closable if it satisfies either of the following equivalent conditions: (a) there exists a closed operator $T$ such that $S \subseteq T$; (b) if $\xi \neq 0$, then $(0,\xi)$ does not belong to the closure of $G(S)$. It is clear that a closable operator admits a smallest closed extension $\overline{S}$ which is characterized by the equation $G(\overline{S}) = \overline{G(S)}$.

It is a standard fact that if $S$ is a densely defined operator, then $G(S^*) = \{(-S\xi,\xi): \xi \in \text{dom } S\}^{\perp}$ (in $\mathcal{H} \oplus \mathcal{H}$). A consequence of this fact is that a densely defined operator $S$ is closable if and only if $S^*$ is densely defined, in which case $\overline{S} = S^{**}$.

The operator $H$ is said to be self-adjoint if $H$ is densely defined and $H = H^*$. The spectral theorem extends to unbounded self-adjoint operators. Recall that a scalar $\lambda$ belongs to the spectrum sp $H$ if and only if it is not the case that $\text{ran}(H - \lambda) = \mathcal{H}$, $\ker(H - \lambda) = \{0\}$ and $(H - \lambda)^{-1}$ is a bounded operator on $\mathcal{H}$. The formulation of the spectral theorem for unbounded $H$ is as follows: if $H = H^*$, then sp $H \subseteq \mathbb{R}$ and there exists a spectral measure $e(\cdot)$ defined on sp $H$ such that $\xi \in$ dom $H$ if and only if $\int|\lambda|^2 d\mu_{\xi,\xi}(\lambda) < \infty$ (with $\mu_{\xi,\xi}$ as before) in which case $\langle H\xi,\eta\rangle = \int \lambda \, d\mu_{\xi,\eta}(\lambda)$ for all $\eta$ in $\mathcal{H}$. As in the bounded case, there is a "functional calculus" ($f \to f(H)$) for $H$.

The polar decomposition theorem also extends to closed densely defined operators: every closed densely defined operator $S$ admits a decomposition $S = uH$ which is uniquely determined by the conditions: $u$ is a partial isometry and $H$ is a positive self-adjoint operator with domain equal to dom $S$ satisfying $\ker S = \ker u = \ker H$. As in the bounded case, $H$ is the unique positive self-adjoint operator satisfying $H^2 = S^*S$.

In the second chapter, we would need to study a conjugate linear operator $S$ (i.e. $S(\lambda\xi + \eta) = \overline{\lambda}S\xi + S\eta$) which is possibly unbounded. In this case, $S^*$ is the unique conjugate linear operator defined on dom $S^* = \{\eta \in \mathcal{H}: \exists\, c > 0 \ni |\langle S\xi,\eta\rangle| \leqslant c\|\xi\| \ \forall \xi \in \text{dom } S\}$ and satisfying $\langle S\xi,\eta\rangle = \langle S^*\eta,\xi\rangle$ for all $\xi \in$ dom $S$ and $\eta \in$ dom $S^*$. (It should have been stated that $S$ was densely defined, for, otherwise, the asserted uniqueness is not valid. However, here and elsewhere, we shall only consider densely defined operators.) The polar decomposition is valid in this context too, with the modification that the "polar part" $u$ will now be a conjugate linear partial isometry. These facts may be derived from the linear case by viewing $\bar{S}$ as a linear operator from dom $S$ into the conjugate Hilbert space $\bar{\mathcal{H}}$.

## 0.2. The Predual $\mathcal{L}(\mathcal{H})_*$

For $\xi,\eta$ in $\mathcal{H}$, let $t_{\xi,\eta}$ be the operator defined by $t_{\xi,\eta}\zeta = \langle\zeta,\eta\rangle\xi$ for $\zeta$ in $\mathcal{H}$. It is an easy consequence of the Cauchy-Schwarz inequality that $\|t_{\xi,\eta}\| = \|\xi\|\ \|\eta\|$; in particular, if $\xi,\eta \neq 0$, then $t_{\xi,\eta}$ is an operator of rank one. Conversely, it is clear that every operator $x$ of rank one is expressible as $\lambda t_{\xi,\eta}$, where $\lambda = \|x\|$ and $\xi$ and $\eta$ are unit vectors. More generally, it can be seen that every operator $x$ of rank $n$ is expressible as

$$x = \sum_{k=1}^{n} \lambda_k t_{\xi_k,\eta_k}, \quad \text{where } \lambda_1 \geq \ldots \geq \lambda_n > 0$$

and $\{\xi_k\}_{k=1}^{n}$ (respectively $\{\eta_k\}_{k=1}^{n}$) is an orthonormal basis of ran $x$ (resp., ran $x^*$).

Recall that an operator $x$ is said to be compact if $\{x\xi: \|\xi\| \leq 1\}$ is contained in a (norm-) compact set. It is easy to see that if $M$ is a closed subspace of $\mathcal{H}$ such that $\|x\xi\| \geq \epsilon\|\xi\|$ for all $\xi$ in $M$ and some $\epsilon > 0$, and if $x$ is compact, then $M$ is necessarily finite-dimensional. Suppose now that $x$ is a positive compact operator. Applying the last statement to ran $1_{[\epsilon,\infty)}(x)$, it follows that ran $1_{[\epsilon,\infty)}(x)$ is finite-dimensional for each $\epsilon > 0$. It is immediate now that sp $x\backslash\{0\}$ is at most countable, and that $x = \Sigma_n \lambda_n t_{\xi_n,\xi_n}$, where $\lambda_n \downarrow 0$ (if there are infinitely many non-zero $\lambda_n$) and $\{\xi_n\}$ is an orthonormal basis for $\overline{\text{ran } x}$ = ran $1_{(0,\infty)}(x)$. (If $x$ has finite rank, the sum is finite, and so we have not specified the range of values of $n$.)

It is easily verified that the set $K(\mathcal{H})$ of compact operators is a norm-closed two-sided ideal in $\mathcal{L}(\mathcal{H})$. (Verify!) If $x \in K(\mathcal{H})$ has polar decomposition $x = u|x|$, then $|x| = u^*x$ is a positive compact operator. So, by the last paragraph $|x| = \Sigma\ \lambda_n t_{\eta_n,\eta_n}$ where $\{\eta_n\}$ is an orthonormal basis for $\overline{\text{ran } |x|}$ = $\ker^{\perp}|x|$.. Since $\ker^{\perp}|x|$ is the initial space of $u$, conclude that if $\xi_n = u\eta_n$, then $\{\xi_n\}$ is an orthonormal basis for $\overline{\text{ran } x}$ such that $x = \Sigma\ \lambda_n\ t_{\xi_n,\eta_n}$. It is clear from the construction that $\|x\| = \|\ |x|\ \| = \sup_n \lambda_n = \lambda_1$.

For arbitrary $\xi,\eta \in \mathcal{H}$, define $\omega_{\xi,\eta} \in K(\mathcal{H})^*$ by $\omega_{\xi,\eta}(x) = \langle x\xi,\eta\rangle$. Then, clearly $\|\omega_{\xi,\eta}\| = \|\xi\|\ \|\eta\|$. (The inequality $\leq$ follows from Cauchy-Schwarz, while the reverse inequality is obtained on considering $x = t_{\eta,\xi}$.) The following exercises lead to an identification of the dual space $K(\mathcal{H})^*$.

**Exercises**

**(0.2.1).** If $\omega \in K(\mathcal{H})^*$, there exists a unique operator $t(\omega) \in \mathcal{L}(\mathcal{H})$ such that $\langle t(\omega)\xi,\eta\rangle = \omega(t_{\xi,\eta})$ for $\xi,\eta \in \mathcal{H}$. Further, $t(\omega)$ is compact and $\Sigma|\langle t(\omega)\xi_i,\xi_i\rangle| < \infty$ for any orthnormal basis $\{\xi_i\}$ of $\mathcal{H}$. (Hint: Consider the bounded sesquilinear form $[\xi,\eta] = \omega(t_{\xi,\eta})$ to lay hands on $t(\omega)$; for a finite orthonormal set $\{\xi_1, ..., \xi_n\}$, let

$$x = \sum_{i=1}^{n} \theta_i t_{\xi_i,\xi_i}$$

where $\theta_i$ are complex numbers of unit modulus such that

$$\theta_i\omega(t_{\xi_i,\xi_i}) = |\omega(t_{\xi_i,\xi_i})|$$

and note that

$$\sum_1^n |\langle t(\omega)\xi_i,\xi_i\rangle| = \omega(x) \leqslant \|\omega\|.$$

For compactness of $t(\omega)$, first reduce to the case $t(\omega) \geqslant 0$ by considering $\phi \in K(\mathcal{H})^*$ given by $\phi(x) = \omega(xu^*)$, where $t(\omega) = uh$ is the polar decomposition of $t(\omega)$ and noting that $t(\phi) = h$; if $t(\omega) \geqslant 0$, use the already established inequality $\Sigma|\langle t(\omega)\xi_i,\xi_i\rangle| \leqslant \|\omega\|$ and conclude that $\text{ran}_{[\epsilon,\infty)}(t(\omega)^{1/2})$ is finite-dimensional for each positive $\epsilon$, and hence that $t(\omega)^{1/2}$ is compact, being a norm-limit of finite-rank operators.)

**(0.2.2).** If $\omega \in K(\mathcal{H})^*$, then $\omega$ admits a decomposition

$$\omega = \sum_{n=1}^{\infty} \alpha_n \omega_{\xi_n,\eta_n},$$

where $\alpha_n \geqslant 0$, $\Sigma\, \alpha_n < \infty$ and $\{\xi_n\}$ and $\{\eta_n\}$ are a pair of orthonormal sequences. Conversely every such system defines an element of $K(\mathcal{H})^*$ as above. Further, $\|\omega\| = \Sigma\, \alpha_n$. (Hint: if $t(\omega) = \Sigma\, \alpha_n t_{\eta_n,\xi_n}$ is the canonical decomposition of the compact operator $t(\omega)$ and if $\beta = (\beta_n) \in c_0$ with $\beta_n \geqslant 0$, observe that

$$x = \Sigma\, \beta_n t_{\xi_n,\eta_n} \in K(\mathcal{H})$$

and that

$$\Sigma\, \alpha_n\beta_n = \omega(x) \leqslant \|\omega\|\, \|x\| = \|\omega\|\, \|\beta\|_{c_0},$$

and appeal to the classical result $c_0^* = \ell^1$, to conclude that $\Sigma\, \alpha_n \leqslant \|\omega\|$.) □

Let $\mathcal{L}(\mathcal{H})_* = \{t(\omega): \omega \in K(\mathcal{H})^*\}$. Note that $\omega \to t(\omega)$ is a bijection between $K(\mathcal{H})^*$ and $\mathcal{L}(\mathcal{H})_*$. Hence $\mathcal{L}(\mathcal{H})_*$ is a Banach space when normed thus: $\|t(\omega)\|_1 = \|\omega\|$. As noted in Ex. (0.2.1), $\mathcal{L}(\mathcal{H})_* \subseteq K(\mathcal{H})$.

Operators in $\mathcal{L}(\mathcal{H})_*$ are called trace-class operators. The notation $\mathcal{L}(\mathcal{H})_*$ is vindicated by the next set of exercises which establish that the dual space $(\mathcal{L}(\mathcal{H})_*)^*$ is isometrically isomorphic to $\mathcal{L}(\mathcal{H})$. For this reason, $\mathcal{L}(\mathcal{H})_*$ is also referred to as the predual of $\mathcal{L}(\mathcal{H})$.

**Exercises**

**(0.2.3).** Any $\rho \in \mathcal{L}(\mathcal{H})_*$ admits a decomposition

$$\rho = \sum_{n=1}^{\infty} \alpha_n t_{\xi_n,\eta_n},$$

where $\alpha_n \geq 0$, $\Sigma\, \alpha_n < \infty$ and $\{\xi_n\}$ and $\{\eta_n\}$ are a pair of orthonormal sequences. Conversely, every such system defines an element $\rho$ of $\mathcal{L}(\mathcal{H})_*$ as above. Further $\|\rho\|_{\mathcal{L}(\mathcal{H})_*} = \Sigma\alpha_n$. (Hint: Note that $t_{\xi_n,\eta_n} = t(\omega_{\xi_n,\eta_n})$ and appeal to Ex. (0.2.2).)

**(0.2.4).** If $x \in \mathcal{L}(\mathcal{H})$, there exists a unique $\phi_x$ in $(\mathcal{L}(\mathcal{H})_*)^*$ satisfying $\phi_x(t_{\xi,\eta}) = \langle x\xi,\eta\rangle$ for all $\xi,\eta$ in $\mathcal{H}$. The map $x \to \phi_x$ defines an isometric isomorphism of $\mathcal{L}(\mathcal{H})$ onto $(\mathcal{L}(\mathcal{H})_*)^*$. □

Since $K(\mathcal{H})$ is a two-sided ideal in $\mathcal{L}(\mathcal{H})$, one can define an $\mathcal{L}(\mathcal{H})$-bimodule structure on $K(\mathcal{H})^*$ as follows: if $a \in \mathcal{L}(\mathcal{H})$, $\omega \in K(\mathcal{H})^*$ and $x \in K(\mathcal{H})$, let

(1) $$(a\cdot\omega)(x) = \omega(xa); \quad (\omega\cdot a)(x) = \omega(ax).$$

Verify that with $a,\omega$ as above,

(2) $$t(a\cdot\omega) = at(\omega); \quad t(\omega\cdot a) = t(\omega)a.$$

Thus $\mathcal{L}(\mathcal{H})_*$ is a two-sided ideal in $\mathcal{L}(\mathcal{H})$. For $\rho \in \mathcal{L}(\mathcal{H})_*$, define the trace of $\rho$ by

$$\mathrm{tr}\,\rho = \langle\rho,1\rangle = \phi_1(\rho),$$

in the notation of Ex. (0.2.4). Then, if $a \in \mathcal{L}(\mathcal{H})$, it follows from the equations (2), with $\rho = t(\omega)$, that

$$\mathrm{tr}\,\rho a = \langle\rho a,1\rangle = \langle\rho,a1\rangle = \langle\rho,1a\rangle = \mathrm{tr}\,a\rho.$$

In the sequel, we shall write $\|\cdot\|_1$ for the norm on $\mathcal{L}(\mathcal{H})_*$ and $\|\cdot\|$ will be used only for the norms on $\mathcal{L}(\mathcal{H})$ and $\mathcal{H}$.

**Exercises**

**(0.2.5).** Let $\rho \in \mathcal{L}(\mathcal{H})_*$. Let $\rho = \Sigma\, \alpha_n t_{\xi_n,\eta_n}$ be a decomposition of $\rho$ as in Ex. (0.2.3).

(a) The series $\Sigma\, \alpha_n\, t_{\xi_n,\eta_n}$ is convergent in $\|\cdot\|_1$, and consequently, for any $x$ in $\mathcal{L}(\mathcal{H})$, tr $\rho x = \langle \rho,x\rangle = \Sigma\, \alpha_n \langle x\xi_n,\eta_n\rangle$; in particular, tr $\rho = \Sigma\, \alpha_n \langle \xi_n,\eta_n\rangle$.
(b) Verify that $\rho^*\rho = \Sigma\, \alpha_n^2 t_{\eta_n,\eta_n}$ and hence that $\|\rho\|_1 = \Sigma\, \alpha_n = \mathrm{tr}(\rho^*\rho)^{1/2}$.
(c) Show that tr $\rho = \Sigma \langle \rho\zeta_n,\zeta_n\rangle$ for any orthonormal basis $\{\zeta_n\}$ of $\mathcal{H}$. (Hint: Appeal to Cauchy and Schwarz to justify an appeal to Parseval.) □

The use of the word "trace" is vindicated by part (c) of the last exercise: in any matrix-representation of $\rho$, the main diagonal is a summable sequence (by Ex. (0.2.1)) and the sum, which does not depend on the choice of co-ordinate system, is the trace of $\rho$.

## 0.3. Three Locally Convex Topologies on $\mathcal{L}(\mathcal{H})$

For several reasons, the norm topology is not a very good topology on $\mathcal{L}(\mathcal{H})$. For example, $\mathcal{L}(\mathcal{H})$ is nonseparable. (Reason: represent $\mathcal{H}$ as $L^2[0,1]$, and for $0 \leq t \leq 1$, let $p_t$ be the projection onto the subspace of functions supported in $[0,t]$; if $s < t$, then $p_t - p_s$ is a non-zero projection and hence has norm one.) Also, if $\mathcal{M}_n$ is an increasing sequence of subspaces and $\mathcal{M} = \overline{\cup \mathcal{M}_n}$, if the sequence $\{\mathcal{M}_n\}$ is not eventually constant, it is not true that $p_n \to p$ in the norm. It turns out that, in such situations, one would do better to consider certain other topologies on $\mathcal{L}(\mathcal{H})$.

Let us briefly recall something about locally convex topologies on vector spaces. Recall that a seminorm on a vector space $M$ is a mapping $p: M \to [0,\infty)$ such that $p(\lambda x) = |\lambda| p(x)$ and $p(x + y) \leq p(x) + p(y)$ whenever $x,y \in M$ and $\lambda \in \mathbb{C}$. Suppose that a family $\{p_i: i \in I\}$ of seminorms on $M$ is given. The induced topology on $M$ is the smallest vector topology on $M$ with respect to which each $p_i$ is continuous at the origin; in this topology, a net $\{x_\alpha: \alpha \in \Lambda\}$ in $M$ converges to $x$ in $M$ if and only if $p_i(x_\alpha - x) \to 0$ for each $i \in I$.

**Definition 0.3.1.** (a) The strong topology on $\mathcal{L}(\mathcal{H})$ is the topology induced by the family of seminorms $\{p_\xi: \xi \in \mathcal{H}\}$ defined by $p_\xi(x) = \|x\xi\|$.

(b) The weak topology on $\mathcal{L}(\mathcal{H})$ is the topology induced by the family of seminorms $\{p_{\xi,\eta}: \xi,\eta \in \mathcal{H}\}$ defined by $p_{\xi,\eta}(x) = |\langle x\xi,\eta\rangle|$.

(c) The $\sigma$-weak topology on $\mathcal{L}(\mathcal{H})$ is the topology induced by the family of seminorms $\{p_\rho: \rho \in \mathcal{L}(\mathcal{H})_*\}$ defined by $p_\rho(x) = |\mathrm{tr}\; x\rho|$. □

In simpler language, a net $\{x_i\}$ in $\mathcal{L}(\mathcal{H})$ converges to $x$ with respect to the:

(a) strong topology iff $\|x_i\xi - x\xi\| \to 0$ for all $\xi \in \mathcal{H}$, i.e., iff $x_i\xi \to x\xi$ in the strong topology on $\mathcal{H}$ ;

(b) weak topology iff $|\langle x_i\xi,\eta\rangle - \langle x\xi,\eta\rangle| \to 0$ for $\xi,\eta$ in $\mathcal{H}$, i.e., iff $x_i\xi \to x\xi$ in the weak topology on $\mathcal{H}$, for every $\xi$ in $\mathcal{H}$ ;
(c) $\sigma$-weak topology iff $|\mathrm{tr}\ \rho(x_i - x)| \to 0$ for every $\rho$ in $\mathfrak{L}(\mathcal{H})_*$, or, equivalently, iff $x_i \to x$ in the weak* topology inherited by $\mathfrak{L}(\mathcal{H})$, by virtue of its being the dual space of $\mathfrak{L}(\mathcal{H})_*$.

Some elementary features of these topologies are listed in the exercises; also included in the exercises are the definitions of three other topologies, namely, the $\sigma$-strong, strong-* and $\sigma$-strong* topologies.

## Exercises

**(0.3.1).**

(a) Show that a net $\{x_i\}$ converges to $x$ weakly if and only if $\mathrm{tr}\ \rho x_i \to \mathrm{tr}\ \rho x$ for every operator $\rho$ of finite rank.
(b) If $S$ is a norm-bounded set in $\mathfrak{L}(\mathcal{H})$ (i.e., $\sup\{\|x\|: x \in S\} < \infty$), show that the weak and $\sigma$-weak topologies, when restricted to $S$, coincide. (Hint: Use (a) and the fact that finite-rank operators are dense in $\mathfrak{L}(\mathcal{H})_*$.)

**(0.3.2).**

(a) When $\mathfrak{L}(\mathcal{H})$ is equipped with either the weak or the $\sigma$-weak topology, the adjoint operation $x \to x^*$ is a continuous self-map.
(b) Let $\mathcal{H}$ be infinite-dimensional. Then show that the adjoint map is not continuous with respect to strong topology. (Hint: let $u$ be a unilateral shift; i.e.,

$$u = \sum_{n=1}^{\infty} t_{\xi_{n+1},\xi_n},$$

where $\{\xi_n\}_{n=1}^{\infty}$ is an orthonormal basis; verify that $u^{*n} \to 0$ strongly while $u^n \not\to 0$ strongly, as $u^n$ is an isometry for each $n$.)

**(0.3.3).**

(a) When $\mathcal{H}$ is infinite-dimensional, show that $\{x \in \mathfrak{L}(\mathcal{H}): x^2 = 0\}$ is strongly dense in $\mathfrak{L}(\mathcal{H})$. (Hint: let $a \in \mathfrak{L}(H)$; a typical basic neighbourhood of $a$ in the strong topology is of the form $\{x \in \mathfrak{L}(\mathcal{H}): \|(x - a)\xi_i\| < \epsilon$ for $1 \leq i \leq n\}$, for some set $\{\xi_1, ..., \xi_n\} \subseteq \mathcal{H}$ and $\epsilon > 0$. Argue that the $\xi_i$'s may, without loss of generality, be assumed to be linearly independent, even orthonormal. Then pick $\eta_1, ..., \eta_n \in \mathcal{H}$ satisfying (i) $\|a\xi_i - \eta_i\| < \epsilon$ for each $i$, and (ii) $\{\xi_1, ..., \xi_n, \eta_1, ..., \eta_n\}$ is linearly independent. Let $x$ be a finite rank operator such that $x\xi_i = \eta_i$ and $x\eta_i = 0$ for each $i$ and $x\zeta = 0$ whenever $\zeta \in \{\xi_1, ..., \xi_n, \eta_1, ..., \eta_n\}^{\perp}$.)

(b) When $\mathcal{H}$ is infinite-dimensional, show that multiplication is not jointly continuous when $\mathcal{L}(\mathcal{H})$ is equipped with any of the three topologies: strong, weak, $\sigma$-weak. (Hint: For the strong topology, use (a); for the $\sigma$-weak and weak topologies, with $u$ as in Ex. (0.3.2)(b), note that $u^{*n} \to 0$ $\sigma$-weakly and hence weakly, so $u^n \to 0$ $\sigma$-weakly and weakly; however, $u^{*n}u^n = 1$ for all $n$.)

(c) If $S$ is a norm-bounded subset of $\mathcal{L}(\mathcal{H})$, then multiplication is jointly continuous, when restricted to $S$, with respect to the strong topology. (The example in (b) shows that even this is false in the $\sigma$-weak and weak topologies).

(d) In all three topologies, show that multiplication is separately continuous; i.e., if $x \in \mathcal{L}(H)$ and $y_i \to y$, then $xy_i \to xy$ and $y_ix \to yx$.

**(0.3.4).** A net $\{x_i\}$ in $\mathcal{L}(\mathcal{H})$ is said to converge to $x$ in $\mathcal{L}(\mathcal{H})$ with respect to the:

(i) $\sigma$-strong topology $\Leftrightarrow \sum_n \|(x_i - x)\xi_n\|^2 \to 0$ whenever $\sum_n \|\xi_n\|^2 < \infty$;

(ii) $\sigma$-strong* topology $\Leftrightarrow \sum_n (\|(x_i - x)\xi_n\|^2 + \|(x_i - x)^*\xi_n\|^2) \to 0$ whenever $\sum_n \|\xi_n\|^2 < \infty$;

(iii) strong* topology $\Leftrightarrow (\|(x_i - x)\xi\|^2 + \|(x_i - x)^*\xi\|^2) \to 0$ for each $\xi$ in $\mathcal{H}$.

(a) Show that the adjoint operation is continuous in the strong* and $\sigma$-strong* topologies, but not in the $\sigma$-strong topology.

(b) Show that, when restricted to norm-bounded sets, multiplication is jointly continuous with respect to each of the three topologies defined in this problem.

(c) Let $\{x_i\}$ be a net in $\mathcal{L}(\mathcal{H})$. Prove the following:

$$x_i \to 0 \quad \text{(strongly)} \quad \Leftrightarrow x_i^*x_i \to 0 \quad \text{(weakly)}$$

$$x_i \to 0 \quad (\sigma\text{-strongly}) \quad \Leftrightarrow x_i^*x_i \to 0 \quad (\sigma\text{-weakly})$$

$$x_i \to 0 \quad \text{(strongly*)} \quad \Leftrightarrow x_i^*x_i + x_ix_i^* \to 0 \quad \text{(weakly)}$$

$$x_i \to 0 \quad (\sigma\text{-strongly*}) \quad \Leftrightarrow x_i^*x_i + x_ix_i^* \to 0 \quad (\sigma\text{-weakly}).$$

(d) When restricted to norm-bounded sets, show that the $\sigma$-strong (resp., $\sigma$-strong*) topology coincides with the strong (resp., strong*) topology.

**(0.3.5).** (a) If $\tau_1$ and $\tau_2$ are topologies on a set $X$, show that the following conditions are equivalent:

(i) $id_X: (X,\tau_2) \to (X,\tau_1)$ is continuous;

(ii) $x_i \to x(\tau_2) \Rightarrow x_i \to x(\tau_1)$, for any net $\{x_i\}$ in $X$ and $x$ in $X$;
(iii) every $\tau_1$-closed set is $\tau_2$-closed. When these conditions hold, write $\tau_1 \subseteq \tau_2$.

(b) Prove the following "inclusion" relations between the different topologies on $\mathfrak{L}(\mathcal{H})$:

$$\begin{array}{ccccccc} \text{Norm} \supseteq & \sigma\text{-strong*} & \supseteq & \sigma\text{-strong} & \supseteq & \sigma\text{-weak} \\ & \cup\| & & \cup\| & & \cup\| \\ & \text{strong} & \supseteq & \text{strong} & \supseteq & \text{weak.} \end{array}$$

(c) Prove, by examples, that if $\mathcal{H}$ is infinite-dimensional, each of the above inclusions is strict.

**(0.3.6).** Let $\{\eta_m\}_{m=1}^{\infty}$ be an orthonormal basis for $\mathcal{H}$, and let $x_m = m^{1/2}t_{\eta_m,\eta_m}$ for each $m$.

(a) Show that 0 belongs to the $\sigma$-strong* closure of $\{x_m\}_{m=1}^{\infty}$, (Hint: if $\{\xi_n\}_{n=1}^{\infty}$ satisfies

$$\sum_n \|\xi_n\|^2 = \sum_{n,m} |<\xi_n,\eta_m>|^2 < \infty,$$

then

$$\sum_n |<\xi_n,\eta_m>|^2 \leqslant m^{-1}\epsilon$$

for infinitely many $m$, for any $\epsilon > 0$, since the harmonic series diverges.)
(b) Show that no subsequence of $\{x_m\}$ converges weakly to zero. (Hint: Uniform boundedness principle!)
(c) $\mathfrak{L}(\mathcal{H})$ is not metrizable -- in fact, does not even satisfy the first axiom of countability -- with respect to any of the topologies mentioned in Ex. (0.3.5) (b) except the norm topology. (Thus, sequences will not do; it is necessary to consider nets.)
(d) Show that on norm-bounded sets, each of the six topologies ($\sigma$-strong* to weak) is metrizable. (Hint: use a dense sequence in the unit ball of $\mathcal{H}$; the assumption of separability is still in force.) □

## 0.4. The Double Commutant Theorem

The commutant of a subset $S$ of $\mathfrak{L}(\mathcal{H})$ is defined to be

$$S' = \{x' \in \mathfrak{L}(\mathcal{H}) : x'x = xx' \text{ for all } x \in S\}.$$

More generally, write $S'' = (S')'$ and $S^{(n+1)} = (S^{(n)})'$.

**Proposition 0.4.1.** *Let* $S, T \subseteq \mathcal{L}(\mathcal{H})$.

(a) $S \subseteq T \Rightarrow T' \subseteq S'$;
(b) $S \subseteq S'' = S^{(2n)}$ *and* $S' = S^{(2n-1)}$ *for* $n \geq 1$;
(c) $S$ *is self-adjoint* $\Rightarrow S'$ *is self-adjoint.*
(d) $S'$ *is, for any* $S$, *a weakly closed subalgebra of* $\mathcal{L}(\mathcal{H})$ *and* $1 \in S'$.

**Proof.** Exercise! □

Before proceeding further, it would help to set up some notation and terminology. For a subset $S$ of $\mathcal{H}$, we shall always write $[S]$ for the smallest closed subspace of $\mathcal{H}$ which contains $S$; for $S \subseteq \mathcal{L}(\mathcal{H})$ and $S \subseteq \mathcal{H}$, we shall simply write $SS$ for $\{x\xi : x \in S, \xi \in S\}$. A set $S$ of operators on $\mathcal{H}$ is said to be non-degenerate if $[S\mathcal{H}] = \mathcal{H}$. Since $\text{ran}^{\perp} x = \ker x^*$, it follows that if $S$ is self-adjoint, then $S$ is non-degenerate if and only if $S\xi = \{0\}$ implies $\xi = 0$.

The stage is now set for von Neumann's double commutant theorem, whose power will be illustrated in the rest of this section.

**Theorem 0.4.2.** *Let* $M$ *be a non-degenerate self-adjoint algebra of operators on* $\mathcal{H}$. *The following conditions on* $M$ *are equivalent:*

(i) $M = M''$.
(ii) $M$ *is weakly closed.*
(iii) $M$ *is strongly closed.*

**Proof.** The implications (i) $\Rightarrow$ (ii) $\Rightarrow$ (iii) are immediate (cf. Ex. (0.3.5)(a), (b)). To prove (iii) $\Rightarrow$ (i), it clearly suffices to prove the following:

(*) If $a'' \in M''$, $\xi_1, \ldots, \xi_n \in \mathcal{H}$ and $\epsilon > 0$, there exists $a \in M$ such that $\|(a'' - a)\xi_i\| < \epsilon$ for $1 \leq i \leq n$.

We first verify (*) in case $n = 1$. Let $\mathcal{M} = [M\xi_1]$ and $p' = p_{\mathcal{M}}$. It is clear that $M\mathcal{M} \subseteq \mathcal{M}$, and so $p'xp' = xp'$ for all $x$ in $M$. Since $M$ is self-adjoint, if $x \in M$, then $x^* \in M$ and so $p'x^*p' = x^*p'$. Comparison of the adjoint of this equation with the previous equation yields $p'x = xp'$ for all $x$ in $M$, whence $p' \in M'$. So $a''p' = p'a''$, and hence $a''\mathcal{M} \subseteq \mathcal{M}$. Since $M\xi_1$ is dense in $\mathcal{M}$, it suffices to prove that $\xi_1 \in \mathcal{M}$. For any $x$ in $M$, clearly $p'x\xi_1 = x\xi_1$ and so $x(1-p')\xi_1 = x\xi_1 - xp'\xi_1 = x\xi_1 - p'x\xi_1 = 0$; thus $M(1 - p')\xi_1) = 0$. The assumed non-degeneracy (and self-adjointness) of $M$ ensures that $(1 - p')\xi_1 = 0$, and hence $\xi_1 \in \mathcal{M}$.

Returning to (*) for general $n$, let $\tilde{\mathcal{H}}$ be the direct sum of $n$ copies of $\mathcal{H}$. Every operator on $\tilde{\mathcal{H}}$ corresponds naturally to an $(n \times n)$ $\mathcal{L}(\mathcal{H})$-valued matrix, this correspondence being a *-algebra isomorphism. With this identification, let

$$\tilde{M} = \left\{ \begin{bmatrix} a & 0 & \cdots & 0 \\ 0 & a & \cdots & 0 \\ \vdots & \vdots & \ddots & \vdots \\ 0 & 0 & \cdots & a \end{bmatrix} : a \in M \right\}.$$

It is relatively painless to verify that (a) $\tilde{M}$ is a non-degenerate self-adjoint algebra of operators on $\mathcal{H}$; (b) $\tilde{M}' = \{((a'_{ij})): a'_{ij} \in M' \; \forall i,j\}$; and

$$\text{(c)} \qquad \tilde{M}'' = \left\{ \begin{bmatrix} a'' & 0 & \cdots & 0 \\ 0 & a'' & \cdots & 0 \\ \vdots & \vdots & \ddots & \vdots \\ 0 & 0 & \cdots & a'' \end{bmatrix} : a'' \in M'' \right\}.$$

Appeal now to the already established case $n = 1$ of (*) with $M$, $\xi_1$ and $\epsilon$ replaced by $\tilde{M}, \xi_1 \oplus \cdots \oplus \xi_n$ and $\epsilon > 0$, to complete the proof of (*) and hence of the theorem. □

Remark that the assertion (*) in the proof of the theorem actually states that if $M$ is a non-degenerate self-adjoint algebra of operators, then $M''$ is (contained in, and hence) equal to the strong closure of $M$. So, if $A$ is any non-degenerate self-adjoint algebra of operators, then $A''$ is the smallest strongly closed (also, weakly closed) self-adjoint algebra containing $A$. Thus, if $S$ is any self-adjoint set of operators, then $S''$ is the strong closure of the algebra generated by $S \cup \{1\}$. A special case is worthy of mention: if $x = x^*$, then $\{x\}''$ is the strong closure of the set of polynomials in $x$, and, in particular, abelian.

**Definition 0.4.3.** A self-adjoint subalgebra $M$ of $\mathcal{L}(\mathcal{H})$ is called a von Neumann algebra if it satisfies $M = M''$. □

By Proposition 0.4.1(d), a von Neumann algebra is a weakly closed self-adjoint unital subalgebra of $\mathcal{L}(\mathcal{H})$, while, by Theorem 0.4.2, any such collection of operators is a von Neumann algebra. Some authors do not require von Neumann algebras to be non-degenerate; they define a von Neumann algebra to be a weakly closed self-adjoint algebra of operators. The difference is inessential and spelt out in the following exercise.

## Exercises

**(0.4.4).** Let $M$ be a (not necessarily non-degenerate) self-adjoint subalgebra of $\mathcal{L}(\mathcal{H})$. Let $e = p_{\mathcal{M}}$, where $\mathcal{M} = [M\mathcal{H}] = (\cap \{\ker x : x \in M\})^{\perp}$. Prove that:

(a) $x = exe$ for all $x$ in $M$; in particular, $\mathcal{M}\mathcal{M} \subseteq \mathcal{M}$;
(b) if $M_e = \{x|\mathcal{M}: x \in M\}$, then $M_e$ is a non-degenerate self-adjoint subalgebra of $\mathcal{L}(\mathcal{M})$;

(c) $M' = \{x' \oplus y \ : x' \in M_e', \ y \in \mathcal{L}(\mathcal{M}^{\perp})\}$, and

$$M'' = \{x'' \oplus \lambda 1_{\mathcal{M}^{\perp}}: x'' \in M_e'', \ \lambda \in \mathbb{C}\}.$$

(Thus, a degenerate von Neumann algebra, as considered by other authors, is just a von Neumann algebra -- in our sense -- of operators on a subspace.)

**(0.4.5).** Let $(X,\mathcal{F},\mu)$ be a separable $\sigma$-finite measure space (so that $L^2(X,\mu)$ is a separable Hilbert space). For $\phi$ in $L^\infty(X,\mu)$, let $m_\phi$ denote the associated multiplication operator: $(m_\phi\xi)(s) = \phi(s)\xi(s)$, for $\xi$ in $L^2(X,\mu) = \mathcal{H}$.

(a) The map $\phi \to m_\phi$ is an isometric* - isomorphism of $L^\infty(X,\mu)$ into $\mathcal{L}(\mathcal{H})$ (where the '*' refers to the assertion $m_\phi^* = m_{\bar{\phi}}$).
(b) If $M = \{m_\phi; \phi \in L^\infty(X,\mu)\}$, then $M = M'$ and consequently $M$ is an abelian von Neumann algebra. (Hint: First, consider the case of finite $\mu$; if $x' \in M'$, show that $x' = m_\phi$ where $\phi = x'\xi_0$, $\xi_0$ being the constant function 1; the general case follows by decomposing $X$ into sets of finite measure. Is $\sigma$-finiteness necessary?)
(c) The $\sigma$-weak and weak topologies on $M$ coincide; under the identification $m_\phi \leftrightarrow \phi$, this topology coincides with the weak* topology inherited by $L^\infty(X,\mu)$ by virtue of its being the dual space of $L^1(X,\mu)$.
(d) A general von Neumann algebra $M$ satisfies $M = M'$ if and only if $M$ is a maximal abelian von Neumann algebra in $\mathcal{L}(\mathcal{H})$.

**(0.4.6).** If $M$ is a von Neumann algebra of operators on $\mathcal{H}$, let $M_\perp = \{\rho \in \mathcal{L}(\mathcal{H})_*: \operatorname{tr} \rho x = 0 \ \forall x \text{ in } M\}$. Then $M_\perp$ is a closed subspace of $\mathcal{L}(\mathcal{H})_*$, $(\mathcal{L}(\mathcal{H})_*/M_\perp)^* \cong M$, and the induced weak* topology on $M$ agrees with the restriction to $M$ of the $\sigma$-weak topology. □

The last exercise shows that every von Neumann algebra admits a predual. It can be shown that such a predual is uniquely determined up to isometric isomorphism, but we shall not go into a proof of that here. Consequently, we may talk of 'the' predual of $M$, which will usually be denoted by $M_*$.

Just as $L^\infty(X,\mu)$ is generated (as a norm-closed subspace) by indicator functions, it is true that every von Neumann algebra $M$ is generated (as a norm-closed subspace) by the set $P(M)$ of its projections. To obtain this and other consequences of the double commutant theorem, it helps to establish a useful preliminary lemma.

Recall that a $C^*$-algebra of operators on $\mathcal{H}$ is a norm-closed self-adjoint subalgebra of $\mathcal{L}(\mathcal{H})$. Clearly von Neumann algebras are

C*-algebras, but the converse is seldom true. (For example, in the notation of Ex. (0.4.5), with $X = [0,1]$ and $\mu$ Lebesgue measure, the set $\{m_\phi: \phi \in C[0,1]\}$ is a proper C*-subalgebra of $M$ which is $\sigma$-weaky dense in $M$.) Now for the lemma.

**Lemma 0.4.7.** *Let $A \subseteq \mathcal{L}(\mathcal{H})$ be a unital C*-algebra. Every element of $A$ is expressible as a linear combination of four unitary operators in $A$.*

**Proof.** Any $x$ in $A$ admits the Cartesian decomposition $x = x_1 + ix_2$, where $x_1$, $x_2$ are self-adjoint operators in $A$. It suffices now to notice that if $x = x^* \in A$ and $\|x\| \leqslant 1$, then

$$x = \frac{1}{2}[\{x + i(1 - x^2)^{1/2}\} + \{x - i(1 - x^2)^{1/2}\}]$$

is an expression of $x$ as an average of two unitary operators; these operators, being continuous functions of $x$, belong to $C^*(x)$, the $C^*$-algebra generated by $x$ and 1, and hence belong to $A$. □

The double commutant theorem, when coupled with the above lemma (applied with $A = M'$), yields the following useful criterion for determining when an operator belongs to a von Neumann algebra.

**Scholium 0.4.8.** *Let $x \in \mathcal{L}(\mathcal{H})$ and $M$ a von Neumann algebra of operators on $\mathcal{H}$. A necessary and sufficient condition for $x$ to belong to $M$ is that $u'xu'^* = x$ for every unitary operator $u'$ in $M'$.*

**Proof.** Exercise! □

**Corollary 0.4.9.** *Let $M$ be a von Neumann algebra and $x \in M$.*

(a) *If $x = u|x|$ is the polar decomposition of $x$, then $u, |x| \in M$;*
(b) *If $x$ is normal, then $1_F(x) \in M$ for every Borel subset $F$ of* sp $x$.

**Proof.** (a) If $u'$ is a unitary operator in $M'$, then $u'xu'^{-1} = x = u|x|$; on the other hand, it is clear that $u'xu'^{-1} = (u'uu'^{-1})(u'|x|u'^{-1})$ is also a (and hence the) polar decomposition of $u'xu'^{-1}$. Hence $u'uu'^{-1} = u$ and $u'|x|u'^{-1} = |x|$. Since $u'$ was arbitrary, the scholium completes the proof.

(b) The uniqueness of the spectral resolution of a normal operator and an argument exactly analogous to the one used in (a) serve to settle this assertion. □

Thus, the scholium implies that just about any canonical construction applied to elements of a von Neumann algebra never leads outside the algebra. It follows from the above Corollary that any von Neumann algebra is generated as a norm-closed subspace by the set of its projections. (Reason: let $M_0$ be the norm closure of the set of linear combinations of projections in $M$; since $M_0$ is self-adjoint, it suffices to verify that if $x = x^* \in M$, then $x \in M_0$; for this, let $\phi_n$ be a sequence of simple functions on sp $x$ such that $\phi_n(t) \to t$ uniformly on sp $x$, and note that by Corollary 0.4.9(b), $\phi_n(x) \in M_0$ for each $n$ and $\lim \|\phi_n(x) - x\| = 0$.)

Before discussing some further properties of a von Neumann algebra, let us briefly digress with some notational conventions. If $\{e_i: i \in I\}$ is any family of projections in a Hilbert space, the symbols $\bigvee_{i\in I} e_i$ and $\bigwedge_{i\in I} e_i$ will denote, respectively, the projections onto the subspaces $[\bigcup_{i\in I} \text{ran } e_i]$ and $\bigcap_{i\in I} \text{ran } e_i$. For a finite collection $e_1, \ldots, e_n$, we shall also write $e_1 \vee \ldots \vee e_n$ and $e_1 \wedge \ldots \wedge e_n$.

## Exercises

**(0.4.10).** If $M$ is a von Neumann algebra and $\{e_i\} \subseteq P(M)$, then $\vee e_i$, $\wedge e_i \in P(M)$. (Thus $P(M)$ has the structure of a complete lattice.) □

An extension of the above exercise is given by the following assertion:

**Proposition 0.4.11.** *Every uniformly bounded monotone (increasing or decreasing) net of self-adjoint operators on $\mathcal{H}$ is weakly convergent.*

**Proof.** Suppose $\{x_i: i \in I\}$ is a net of self-adjoint operators on $\mathcal{H}$ satisfying (a) if $i,j \in I$ and $i \leqslant j$, then $x_i \leqslant x_j$; and (b) there exists a constant $c > 0$ such that $\|x_i\| \leqslant c$ for all $i$ in $I$. For a unit vector $\xi$ in $\mathcal{H}$, $\{\langle x_i\xi,\xi\rangle : i \in I\}$ is a monotone increasing net of real numbers in $[-c,c]$, and consequently convergent to its supremum. It follows from the polarization identity that (cf. Ex. (0.4.12)) for any $\xi,\eta \in \mathcal{H}$, the net $\{\langle x_i\xi,\eta\rangle: i \in I\}$ is convergent. Denoting this limit by $[\xi,\eta]$ it is clear that $[\cdot,\cdot]$ is a bounded (by $c$) sesquilinear form on $\mathcal{H}$. Hence there exists $x$ in $\mathcal{L}(\mathcal{H})$ such that $\langle x\xi,\eta\rangle = [\xi,\eta]$ for all $\xi,\eta \in \mathcal{H}$. Clearly, then, the net $\{x_i: i \in I\}$ converges weakly to $x$.

## Exercises

**(0.4.12).** If $[\cdot,\cdot]$ is a sesquilinear form on a complex vector space $V$, then, for any $\xi,\eta$ in $V$,

$$4[\xi,\eta] = \sum_{k=0}^{3} i^k[\xi + i^k\eta,\ \xi + i^k\eta].$$

**(0.4.13).** Let $\{x_i: i \in I\}$ be a monotone increasing net of self-adjoint operators on $\mathcal{H}$ and let $x = \lim x_i$ (as in Prop. 0.4.11). Then,

(a) $x_i \to x$ strongly. (Hint: if $\|x_i\| \leqslant c$ for all $i$, then $\|(x-x_i)\xi\| \leqslant \|(x-x_i)^{1/2}\| \cdot \|(x-x_i)^{1/2}\xi\| \leqslant (2c)^{1/2}\|(x-x_i)^{1/2}\xi\|$; use Ex. (0.3.4)(c), applied to $\{x_i - x\}$.)

(b) $x_i \to x$ $\sigma$-weakly. (Hint: if $\Sigma_n \|\xi_n\|^2 < \infty$ and $\Sigma_n \|\eta_n\|^2 < \infty$, and $N$ any integer, then

$$\sum_n |\langle (x - x_i)\xi_n, \eta_n \rangle| \leq \sum_{n=1}^{N} |\langle (x - x_i)\xi_n, \eta_n \rangle|$$

$$+ 2c \sum_{n=N+1}^{\infty} \|\xi_n\| \, \|\eta_n\|.)$$

(c) $x_i \to x$ $\sigma$-strongly*. (Since $x_i$, $x$ are self-adjoint, only $\sigma$-strong convergence need be proved; use (a) and Ex. (0.3.4)(d).)

(d) If $y \in \mathcal{L}(\mathcal{H})$ satisfies $x_i \leq y$ for all $i$, then $x \leq y$; for this reason, we shall write $x = \sup x_i$. □

We conclude this section with the definition of a factor and a basic fact concerning factors (Prop. 0.4.17) that will be needed in the next chapter.

**Definition 0.4.14.** Let $M$ be a von Neumann algebra of operators on $\mathcal{H}$.

(a) The set $\{x \in M: xy = yx$ for all $y$ in $M\}$ is called the centre of $M$ and denoted by $Z(M)$.

(b) $M$ is called a factor if $Z(M) = \{\lambda 1: \lambda \in \mathbb{C}\}$.

(c) For a projection $e$ in $M$, the central cover of $e$, denoted by $c(e)$, is the projection defined by $c(e) = \wedge\{f \in \mathcal{P}(M) \cap Z(M): e \leq f\}$. □

It is an easy consequence of the double commutant theorem that the intersection of any family of von Neumann algebras is again a von Neuman algebra; in particular, since $Z(M) = M \cap M'$, the centre of a von Neumann algebra is an abelian Neumann algebra. Hence, by Ex. (0.4.10), $c(e) \in \mathcal{P}(Z(M))$ whenever $e \in \mathcal{P}(M)$; by definition, $c(e)$ is the smallest central projection dominating $e$. The following Exercise leads to a more concrete description of $c(e)$.

## Exercises

(0.4.15).

(a) Let $N$ be a von Neumann algebra of operators on $\mathcal{H}$; let $\mathcal{M}$ be any closed subspace of $\mathcal{H}$ and let $e = p_{\mathcal{M}}$. (It is not assumed that $e \in N$.) Then $\bar{e} = \wedge\{f \in \mathcal{P}(N): e \leq f\}$ is a projection in $N$ and ran $\bar{e}$ $= [N'\mathcal{M}]$, the smallest $N'$-invariant closed subspace containing ran $e$.

(b) Let $N_1$ and $N_2$ be von Neumann algebras acting on $\mathcal{H}$. then $(N_1 \cup N_2)' = N_1' \cap N_2'$ and $(N_1 \cap N_2)' = (N_1' \cup N_2')''$. (Hint: The first assertion is trivial, and implies the second.)

(c) If $M$ is a von Neumann algebra and $e \in \mathcal{P}(M)$, then ran $c(e) =$ $[M\mathcal{M}]$, where $\mathcal{M} = $ ran $e$. □

**Lemma 0.4.16.** *Let $M$ be a von Neumann algebra and $e,f \in P(M)$. The following conditions are equivalent:*

(i) $exf = 0$ *for all* $x$ *in* $M$;
(ii) $c(e)\,c(f) = 0$.

**Proof.** (i) $\Rightarrow$ (ii). The hypothesis is that $M\mathcal{M} \subseteq \ker e$, where $\mathcal{M} = \operatorname{ran} f$. Hence, by Ex. 0.4.15(c), it follows that $\operatorname{ran} c(f) \subseteq \ker e$, whence $ec(f) = 0$. This means $e \leqslant 1 - c(f)$, and so, by the definition of the central cover, $c(e) \leqslant 1 - c(f)$.

(ii) $\Rightarrow$ (i). Reverse the steps of the proof of (i) $\Rightarrow$ (ii). □

**Proposition 0.4.17.** *If $e$ and $f$ are non-zero projections in a factor $M$, there exists a non-zero partial isometry $u$ in $M$ such that $u^*u \leqslant e$ and $uu^* \leqslant f$.*

**Proof.** The assumptions ensure that $c(e) = c(f) = 1$. Lemma 0.4.16 then guarantees the existence of an $x$ in $M$ such that $fxe \neq 0$. Let $fxe = uh$ be the polar decomposition of $fxe$. This $u$ does the job. □

# Chapter 1
# THE MURRAY - VON NEUMANN CLASSIFICATION OF FACTORS

The notion of unitary equivalence, while being most natural, has the disadvantage of not being additive in the following sense: if $e_1$, $e_2$, $f_1$ and $f_2$ are projections such that $e_i$ is unitarily equivalent to $f_i$, for $i$ = 1,2, and if $e_1 \perp e_2$ and $f_1 \perp f_2$, it is not necessarily true that $e_1 + e_2$ is unitarily equivalent to $f_1 + f_2$. This problem disappears if, more generally, one considers two projections as being equivalent if their ranges are the initial and final spaces of a partial isometry. This equivalence, when all the operators concerned -- the projections as well as the partial isometry -- are required to come from a given factor $M$, is the subject of Section 1.1, where the crucial result is that, with respect to a natural order, the set of equivalence classes of the projections in a factor is totally ordered. The next section examines finite projections -- those not equivalent to proper subprojections; the main result being that finiteness is preserved under taking finite suprema. The final section, via a quantitative analysis of the order relation discussed earlier, effects a primary classification of factors into three types. The principal tool used is called a 'relative dimension function' by Murray and von Neumann.

## 1.1. The Relation ... $\sim$ ... (rel *M*)

Henceforth, the symbol $M$ will always denote a von Neumann algebra, and $\mathcal{P}(M)$ the complete lattice of projections in $M$.

**Definition 1.1.1.** Let $e,f \in \mathcal{P}(M)$. We shall write:

(a) $e \sim f$ (rel $M$) or simply $e \sim f$ in case there exists a partial isometry $u$ in $M$ such that $u^*u = e$ and $uu^* = f$;
(b) $e \precsim f$ if there exists $e_1$ in $\mathcal{P}(M)$ such that $e \sim e_1 \leq f$. □

It is readily verified that $\sim$ is indeed an equivalence relation on $\mathcal{P}(M)$ and that the validity of $e \precsim f$ is unimpaired by replacing either $e$ or $f$ by an equivalent projection. We shall adopt the notation $u$: $e \sim f$ to mean that $u$, $e$ and $f$ belong to $M$ and are as in (a) of the above definition.

We shall find it convenient, in this chapter, at least, to work with subspaces rather than projections. Via the transition $\mathcal{M} \to p_{\mathcal{M}}$, we may (and will) use such statements as $u$: $\mathcal{M} \sim \mathcal{M}_1 \subseteq \mathcal{N}$. Since we are only concerned with $\mathcal{P}(M)$, we should only consider subspaces which are the ranges of projections in $M$. It will be useful to consider a slight generalization of this notion.

**Definition 1.1.2.** A (not necessarily closed) linear subspace $\mathcal{D}$ of $\mathcal{H}$ is said to be affiliated to $M$, denoted by $\mathcal{D}\ \eta\ M$, if $a'\mathcal{D} \subseteq \mathcal{D}$ for all $a'$ in $M'$. □

It follows from the double commutant theorem that if $\mathcal{M}$ is a closed subspace, then $\mathcal{M}\eta M$ if and only if $p_{\mathcal{M}} \in M$. In general, there exists several non-closed subspaces affiliated to $M$; if, for instance, there exists $a$ in $M$ such that ran $a$ is not closed, then ran $a$ would be such an example. To deal with such subspaces, it becomes necessary to deal with unbounded operators. In this context, the following definition supplements Definition 1.1.2.

**Definition 1.1.3.** A closed operator $A$ is said to be affiliated to $M$, denoted $A\ \eta\ M$, if $a'A \subseteq Aa'$ for every $a' \in M'$; i.e., if $\xi \in \text{dom } A$ and $a' \in M'$ imply $a'\xi \in \text{dom } A$ and $Aa'\xi = a'A\xi$. □

Observe that for bounded operators, (the double commutant theorem ensures that) the notions 'affiliated to $M$' and 'belonging to $M$' coincide. The following exercises should convince the reader that this notion is a natural one and that it is possible to deal with this notion by considering only bounded operators.

## Exercises

**(1.1.4).** Let $A$ be a closed and densely defined linear operator. The following conditions are equivalent: (i) $A\ \eta\ M$; (ii) $A^*\ \eta\ M$; (iii) if $A = uH$ is the polar decomposition of $A$, then $u \in M$ and $1_F(H) \in M$ for every Borel subset $F$ of $[0,\infty)$.

**(1.1.5).** Let $(X,\mathcal{F},\mu)$ be a separable $\sigma$-finite measure space and $M = \{m_\phi\colon \phi \in L^\infty(X,\mu)\} \subseteq \mathcal{L}(L^2(X,\mu))$ (cf. Ex. (0.4.5)). Show that a closed densely defined operator $A$ on $L^2(X,\mu)$ is affiliated to $M$ if and only if there exists a $\mu$-a.e. finite-valued measurable function $\psi$ such that dom $A = \{\xi \in L^2(X,\mu)\colon \psi\xi \in L^2(X,\mu)\}$ and $A\xi = \psi\xi$ for $\xi$ in dom $A$.

**(1.1.6).** For a closed densely defined operator $A$, let $rp(A)$ (called the range projection of $A$) be the projection onto $\overline{\text{ran } A}$. If $A \,\eta\, M$, show that $rp(A) = 1_{(0,\infty)}(|A^*|) \in M$ and that $rp(A) \sim rp(A^*)$. (Hint: polar decomposition.) □

For the rest of this chapter, the symbols $\mathcal{M}$, $\mathcal{N}$, $\mathcal{B}$ and $\mathcal{R}$ will, unless otherwise specified, always denote closed subspaces affiliated to $M$.

**Proposition 1.1.7.** *The relation ... $\sim$ ... (rel $M$) is countably additive in the following sense: if $\mathcal{M}_n \sim \mathcal{N}_n$ for $n = 1,2, \ldots$, and $\mathcal{M}_m \perp \mathcal{M}_n$ and $\mathcal{N}_m \perp \mathcal{N}_n$ for $m \neq n$, then $\oplus \mathcal{M}_n \sim \oplus \mathcal{N}_n$.*

**Proof.** First observe that $\oplus \mathcal{M}_n, \oplus \mathcal{N}_n \,\eta\, M$ since $\mathcal{M}_n, \mathcal{N}_n \,\eta\, M$. If $u_n: \mathcal{M}_n \sim \mathcal{N}_n$, it is easy to see, under the hypothesis, that the sequence $\{\Sigma_{m=1}^{n} u_m\}_{n=1}^{\infty}$ converges strongly to a partial isometry $u$ such that $u$: $\oplus \mathcal{M}_n \sim \oplus \mathcal{N}_n$. □

**Proposition 1.1.8.** $\mathcal{M} \precsim \mathcal{N}$ *and* $\mathcal{N} \precsim \mathcal{M}$ *imply* $\mathcal{M} \sim \mathcal{N}$.

**Proof.** Let $u$: $\mathcal{M} \sim \mathcal{M}' \subseteq \mathcal{N}$ and $v$: $\mathcal{N} \sim \mathcal{N}_0 \subseteq \mathcal{M}$. Set $\mathcal{M}_1 = w\mathcal{M}$, where $w = vu$. It is clear that $w \in M$ and that $w^n$ is a partial isometry for each $n$ (since $w$ maps $\mathcal{M}$ isometrically into itself) such that $w^n$: $\mathcal{M}_o \sim \mathcal{M}_n$ for each $n \geq 0$, where $\mathcal{M}_0 = \mathcal{M}$ and $\mathcal{M}_n = w^n\mathcal{M}_0$. Since $\mathcal{N}_0 \subseteq \mathcal{M}$, we also have $w^n p_{\mathcal{N}_0}$: $\mathcal{N}_0 \sim \mathcal{N}_n$ for each $n \geq 0$, where $\mathcal{N}_n = w^n \mathcal{N}_0$; consequently, for $n \geq 0$,

$$w^n(p_{\mathcal{M}_0} - p_{\mathcal{N}_0}): (\mathcal{M}_0 \ominus \mathcal{N}_0) \sim (\mathcal{M}_n \ominus \mathcal{N}_n),$$

where we write $\mathcal{M} \ominus \mathcal{N}$ for $\mathcal{M} \cap \mathcal{N}^{\perp}$. The construction shows that $\mathcal{M} = \mathcal{M}_0 \supseteq \mathcal{N}_0 \supseteq \mathcal{M}_1 \supseteq \mathcal{N}_1 \supseteq \ldots$, and so $\cap_n \mathcal{M}_n = \cap_n \mathcal{N}_n = \mathcal{R}$ (say). Appealing to Proposition 1.1.7, we get:

$$\begin{aligned} \mathcal{M} &= \left[\bigoplus_{m=0}^{\infty} (\mathcal{M}_n \ominus \mathcal{N}_n)\right] \oplus \left[\bigoplus_{n=0}^{\infty} (\mathcal{N}_n \ominus \mathcal{M}_{n+1})\right] \oplus \mathcal{R} \\ &\sim \left[\bigoplus_{n=1}^{\infty} (\mathcal{M}_n \ominus \mathcal{N}_n)\right] \oplus \left[\bigoplus_{n=0}^{\infty} (\mathcal{N}_n \ominus \mathcal{M}_{n+1})\right] \oplus \mathcal{R} \\ &= \mathcal{N}_0 \sim \mathcal{N}, \end{aligned}$$

where we have used the fact proved above that $(\mathcal{M}_o \ominus \mathcal{N}_0) \sim (\mathcal{M}_1 \ominus \mathcal{N}_1) \sim (\mathcal{M}_2 \ominus \mathcal{N}_2) \sim \ldots$ . □

**Proposition 1.1.9.** *Suppose $M$ is a factor. If $\mathcal{M}$ and $\mathcal{N}$ are closed subspaces affiliated to $M$, then either $\mathcal{M} \precsim \mathcal{N}$ or $\mathcal{N} \precsim \mathcal{M}$.*

**Proof.** Assume, with no loss of generality, that both $\mathcal{M}$ and $\mathcal{N}$ are non-zero. Let $\mathcal{F}$ denote the set, whose typical member is a family

$\{(\mathcal{M}_i, \mathcal{N}_i): i \in I\}$ where $\mathcal{M}_i, \mathcal{N}_i \neq (0)$ for each $i$, $\mathcal{M}_i \perp \mathcal{M}_j$ and $\mathcal{N}_i \perp \mathcal{N}_j$ for $i \neq j$, $\mathcal{M}_i \sim \mathcal{N}_i$ for all $i$, and $\mathcal{M}_i \subseteq \mathcal{M}$, $\mathcal{N}_i \subseteq \mathcal{N}$ for all $i$. Proposition 0.4.17 ensures that the above set $\mathcal{F}$ is non-void. The set $\mathcal{F}$ is clearly partially ordered by inclusion, and the union of a totally ordered collection of families of the above sort is again a family of the above sort. So, by Zorn's lemma, the set $\mathcal{F}$ has a maximal element, say $\{(\mathcal{M}_i, \mathcal{N}_i): i \in I\}$. Since $\mathcal{H}$ is separable, the index set $I$ is countable. It follows from Proposition 1.1.7 that $\underline{\mathcal{M}} \sim \underline{\mathcal{N}}$ where

$$\underline{\mathcal{M}} = \bigoplus_{i\in I} \mathcal{M}_i \quad \text{and} \quad \underline{\mathcal{N}} = \bigoplus_{i\in I} \mathcal{N}_i .$$

The maximality of the collection $\{(\mathcal{M}_i, \mathcal{N}_i): i \in I\}$ and Proposition 0.4.17 -- applied to $\mathcal{M} \ominus \underline{\mathcal{M}}$ and $\mathcal{N} \ominus \underline{\mathcal{N}}$, in case these are both nonzero -- ensure that $\mathcal{M} = \underline{\mathcal{M}}$ or $\mathcal{N} = \underline{\mathcal{N}}$ ; i.e., $\mathcal{M} \precsim \mathcal{N}$ or $\mathcal{N} \precsim \mathcal{M}$. □

## 1.2. Finite Projections

For the rest of this chapter, the symbol $M$ will always denote a factor of operators on $\mathcal{H}$, the primary reason for this being Proposition 1.1.9.

**Definition 1.2.1.** A projection $e$ in $M$ is said to be finite if $e_0 \in \mathcal{P}(M)$ and $e \sim e_0 \leqslant e$ imply $e_0 = e$. In the contrary case, $e$ is said to be infinite. Correspondingly, a closed subspace $\mathcal{M}$ which is affiliated to $M$ is said to be finite or infinite according as $p_{\mathcal{M}}$ is finite or infinite. □

**Proposition 1.2.2.** *If* $\mathcal{M} \precsim \mathcal{N}$ *and* $\mathcal{N}$ *is finite, then* $\mathcal{M}$ *is finite; in particular, if any infinite* $\mathcal{M}$ *exists, then* $\mathcal{H}$ *is infinite.*

**Proof.** Since finiteness is preserved under equivalence (check it!), assume, with no loss of generality, that $\mathcal{M} \subseteq \mathcal{N}$. If $\mathcal{M} \sim \mathcal{M}_0 \subseteq \mathcal{M}$, then $\mathcal{N} = \mathcal{M} \oplus (\mathcal{N} \ominus \mathcal{M}) \sim \mathcal{M}_0 \oplus (\mathcal{N} \ominus \mathcal{M}) \subseteq \mathcal{N}$ and consequently $(0) = \mathcal{N} \ominus (\mathcal{M}_0 \oplus (\mathcal{N} \ominus \mathcal{M})) = \mathcal{M} \ominus \mathcal{M}_0$, thereby establishing the finiteness of $\mathcal{M}$. In particular, if $\mathcal{H}$ is finite, then every $\mathcal{M}$ is finite, thus establishing the contrapositive of the second assertion. □

The next result is very crucial in the further development of the theory; it is a sort of Euclidean algorithm.

**Proposition 1.2.3.** *Let* $\mathcal{M}, \mathcal{N}\ \eta\ M$ *and* $\mathcal{N} \neq (0)$. *Then there exists a family* $\{\mathcal{N}_i: i \in I\}$ *of pairwise orthogonal subspaces of* $\mathcal{M}$, *and a subspace* $\mathcal{R}$ *of* $\mathcal{M}$ *such that* $\mathcal{N}_i, \mathcal{R}\ \eta\ M$, $\mathcal{M} = (\oplus_{i\in I} \mathcal{N}_i) \oplus \mathcal{R}$, $\mathcal{N}_i \sim \mathcal{N}$ *for all* $i$ *and* $\mathcal{R} \prec \mathcal{N}$ *(in the sense that* $\mathcal{R} \precsim \mathcal{N}$ *but* $\mathcal{R} \not\sim \mathcal{N}$ *).*

*If, in one such decomposition, the index set* $I$ *is infinite, there exists another decomposition in which the remainder term* $\mathcal{R} = (0)$.

**Proof**. If $\mathcal{M} \prec \mathcal{N}$, set $I = \phi$, $\mathcal{R} = \mathcal{M}$. In the alternative case, it follows from Proposition 1.1.9 that $\mathcal{N} \precsim \mathcal{M}$, so that there exists $\mathcal{N}_0 \subseteq \mathcal{M}$ such that $\mathcal{N} \sim \mathcal{N}_0$. As in the proof of Proposition 1.1.9, an appeal to Zorn's lemma yields a family $\{\mathcal{N}_i: i \in I\}$ of pairwise orthogonal subspaces of $\mathcal{M}$, each equivalent to $\mathcal{N}$, with the property that the family is maximal with respect to the above property. If $\mathcal{R} = \mathcal{M} \ominus (\oplus_{i \in I} \mathcal{N}_i)$, the maximality of $\{\mathcal{N}_i\}$ ensures that $\mathcal{R} \not\precsim \mathcal{N}$; consequently, by Proposition 1.1.9, it must be that $\mathcal{R} \prec \mathcal{N}$.

For the second assertion, in view of the separability of $\mathcal{H}$, we may assume that $I = \{1,2,3,...\}$. Then notice that

$$\mathcal{M} = \mathcal{R} \oplus \bigoplus_{n=1}^{\infty} \mathcal{N}_n \sim \mathcal{R} \oplus \bigoplus_{n=2}^{\infty} \mathcal{N}_n \quad (\text{since } \mathcal{N}_n \sim \mathcal{N} \sim \mathcal{N}_{n+1})$$
$$\sim \mathcal{R}' \oplus \bigoplus_{n=2}^{\infty} \mathcal{N}_n,$$

where $\mathcal{R}'$ is a suitable subspace of $\mathcal{N}_1$, since $\mathcal{R} \precsim \mathcal{N} \sim \mathcal{N}_1$; thus,

$$\mathcal{M} \precsim \bigoplus_{n=1}^{\infty} \mathcal{N}_n \subseteq \mathcal{M}.$$

Conclude from Proposition 1.1.8 that $\mathcal{M} \sim \oplus_{n=1}^{\infty} \mathcal{N}_n$. If $u: \mathcal{M} \sim \oplus_{n=1}^{\infty} \mathcal{N}_n$ and $\mathcal{N}_n' = u^* \mathcal{N}_n$, then $\mathcal{M} = \oplus_{n=1}^{\infty} \mathcal{N}_n'$ and $\mathcal{N}_n' \sim \mathcal{N}$ for all $n$. □

**Corollary 1.2.4.** (a) *Let* $(0) \neq \mathcal{M}\ \eta\ M$. *Then* $\mathcal{M}$ *is infinite if and only if* $\mathcal{M}$ *admits a decomposition* $\mathcal{M} = \mathcal{B} \oplus (\mathcal{M} \ominus \mathcal{B})$, *with* $\mathcal{M} \sim \mathcal{B} \sim (\mathcal{M} \ominus \mathcal{B})$.

(b) *If* $\mathcal{M}, \mathcal{N}\ \eta\ M$ *and if* $\mathcal{M}$ *is infinite, then* $\mathcal{N} \precsim \mathcal{M}$; *in particular, any two infinite projections are equivalent.*

**Proof**. One implication in (a) is trivial; so suppose $u: \mathcal{M} \sim \mathcal{M}_1 \subseteq \mathcal{M}$. If $\mathcal{B}_n = u^n(\mathcal{M} \ominus \mathcal{M}_1)$, it is clear that $\mathcal{B}_n \sim \mathcal{B}_1$ for all $n$, that $\mathcal{B}_1 \neq (0)$ and that $\oplus_{n=1}^{\infty} \mathcal{B}_n \subseteq \mathcal{M}$. It follows from the second part of Proposition 1.2.3 that $\mathcal{M}$ admits a decomposition $\mathcal{M} = \oplus_{n=1}^{\infty} \tilde{\mathcal{B}}_n$, where $\tilde{\mathcal{B}}_n \sim \mathcal{B}_1$ for all $n$.

To prove (a), simply put $\mathcal{B} = \oplus_{n=1}^{\infty} \tilde{\mathcal{B}}_{2n}$; for (b), apply Proposition 1.2.3 to $\mathcal{N}$ and $\tilde{\mathcal{B}}_1$ to get $\mathcal{N} = (\oplus_{i \in I} \mathcal{B}_i') \oplus \mathcal{R}$, with $\mathcal{B}_i' \sim \tilde{\mathcal{B}}_1$ and $\mathcal{R} \prec \tilde{\mathcal{B}}_1$. Since $\mathcal{H}$ is separable, the set $I$ is countable; hence

$$\mathcal{N} \precsim \left[\bigoplus_{n=2}^{\infty} \tilde{\mathcal{B}}_n \oplus \mathcal{R}\right] \precsim \left[\bigoplus_{n=2}^{\infty} \tilde{\mathcal{B}}_n \oplus \tilde{\mathcal{B}}_1\right] \subseteq \mathcal{M},$$

as desired. This proves the first half of (b), which, together with the Schröder-Bernstein result (Prop. 1.1.8) establishes the second half of (b). □

The reader might find it pleasantly diverting to read the statement with which Murray and von Neumann preface part (b) of the preceding corollary: 'we are now in the position to determine the chief characteristics of infinity'.

The next few lemmas lead up to a proof of the main result in this section -- that a supremum of two finite projections is again a finite projection. Some of these intermediate results -- particular Lemma 1.2.5 -- are interesting in their own right.

**Lemma 1.2.5.** *Let* $\mathcal{M}$, $\mathcal{N}$, $\mathcal{B}$ $\eta$ $M$, $\mathcal{M} \perp \mathcal{N}$ *and* $\mathcal{B} \subseteq \mathcal{M} \oplus \mathcal{N}$. *Then* $\mathcal{M}$, $\mathcal{N}$ *and* $\mathcal{B}$ *admit decompositions* $\mathcal{M} = \mathcal{M}_1 \oplus \mathcal{M}_2 \oplus \mathcal{M}_3$, $\mathcal{N} = \mathcal{N}_1 \oplus \mathcal{N}_2 \oplus \mathcal{N}_3$ *and* $\mathcal{B} = \mathcal{M}_2 \oplus \mathcal{N}_2 \oplus \mathcal{B}_0$, *with* $\mathcal{M}_i$, $\mathcal{N}_i$, $\mathcal{B}_0$ $\eta$ $M$ *and satisfying*:

$$\mathcal{M}_2 = \mathcal{M} \cap \mathcal{B}, \quad \mathcal{M}_3 = \mathcal{M} \cap \mathcal{B}^{\perp}$$

$$\mathcal{N}_2 = \mathcal{N} \cap \mathcal{B}, \quad \mathcal{N}_3 = \mathcal{N} \cap \mathcal{B}^{\perp}$$

$$\mathcal{M}_1 = \mathcal{M} \ominus (\mathcal{M}_2 \oplus \mathcal{M}_3)$$

$$\mathcal{N}_1 = \mathcal{N} \ominus (\mathcal{N}_2 \oplus \mathcal{N}_3)$$

$$\mathcal{B}_0 = \{\xi + A\xi : \xi \in \text{dom } A\},$$

*where* $A$ *is a closed operator affiliated to* $M$ *such that* $\overline{\text{dom } A} = \mathcal{M}_1$, $\overline{\text{ran } A} = \mathcal{N}_1$, $\ker A = (0)$. *Further* $\mathcal{M}_1 \sim \mathcal{B}_0 \sim \mathcal{N}_1$.

**Proof.** Simply define $\mathcal{M}_1$, $\mathcal{M}_2$, $\mathcal{M}_3$, $\mathcal{N}_1$, $\mathcal{N}_2$, $\mathcal{N}_3$ as above, and let $\mathcal{B}_0 = \mathcal{B} \ominus (\mathcal{M}_2 \oplus \mathcal{N}_2)$. Let $e = p_{\mathcal{M}}$ and $f = p_{\mathcal{N}}$. Note that $\ker(e|\mathcal{B}) = \mathcal{B} \cap \mathcal{N} = \mathcal{N}_2$ and consequently $e|\mathcal{B}_0$ is one-to-one; further

$$\xi \in \mathcal{M} \ominus e\mathcal{B} \Leftrightarrow \xi \in \mathcal{M} \text{ and } 0 = \langle \xi, e\eta \rangle = \langle \xi, \eta \rangle \ \forall \eta \in \mathcal{B}$$

$$\Leftrightarrow \xi \in \mathcal{M} \cap \mathcal{B}^{\perp} = \mathcal{M}_3.$$

It follows that $\overline{e\mathcal{B}} = \mathcal{M}_1 \oplus \mathcal{M}_2$, and hence that $\overline{e\mathcal{B}_0} = \mathcal{M}_1$. Thus $e|\mathcal{B}_0$ maps $\mathcal{B}_0$ one-to-one onto a dense subspace $\mathcal{D}$ of $\mathcal{M}_1$.

An exactly similar reasoning shows that $f|\mathcal{B}_0$ maps $\mathcal{B}_0$ one-to-one onto a dense subspace of $\mathcal{N}_1$. Define $A: \mathcal{D} \to \mathcal{N}_1$ by $A\xi = f\eta$ where $\eta$ is the unique vector in $\mathcal{B}_0$ such that $e\eta = \xi$. It follows at once that $A$ is one-to-one, $\overline{\text{dom } A} = \mathcal{M}_1$, $\overline{\text{ran } A} = \mathcal{N}_1$ and $\mathcal{B}_0 = \{\xi + A\xi : \xi \in \mathcal{D}\}$. The fact that $\mathcal{B}_0$ is a closed subspace of $\mathcal{M}_1 \oplus \mathcal{N}_1$ means precisely

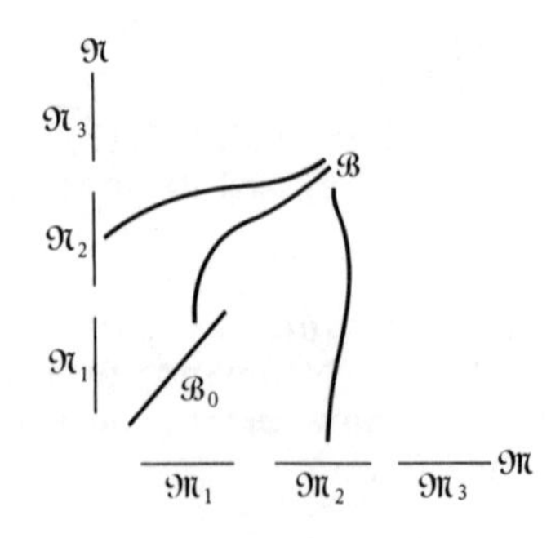

that $A$ is a closed operator. It is a routine matter to verify that $A \,\eta\, M$, since $\mathcal{M}_i$, $\mathcal{N}_i$, $\mathcal{B}_0 \,\eta\, M$.

We already know that ran $\overline{ep_{\mathcal{B}_0}} = \mathcal{M}_1$; if we can show that

$$\overline{\text{ran } p_{\mathcal{B}_0} e} = \mathcal{B}_0,$$

it would follow from Ex. (1.1.6) that $\mathcal{M}_1 \sim \mathcal{B}_0$. The asserted equality can be directly proved without much difficulty; it can also be deduced from the following exercise. Similarly, consideration of $fp_{\mathcal{B}_0}$ would prove $\mathcal{B}_1 \sim \mathcal{B}_0$. □

**Exercise**

**(1.2.6).** Let $\mathcal{H} = \mathcal{M} \oplus \mathcal{N}$, let $\mathcal{D}$ be a dense linear subspace of $\mathcal{M}$ and let $A: \mathcal{D} \to \mathcal{N}$ be a closed operator. Let $\mathcal{B}$ denote the graph of $A$. Then $\mathcal{B}^{\perp} = \{(-A^*\eta,\eta): \eta \in \text{dom } A^*\}$, viewing $A^*$ as a linear operator from the dense subspace dom $A^*$ of $\mathcal{N}$, into $\mathcal{M}$. With respect to the decomposition $\mathcal{H} = \mathcal{M} \oplus \mathcal{N}$, show that

$$p_{\mathcal{M}} = \begin{bmatrix} 1 & 0 \\ 0 & 0 \end{bmatrix}, \quad p_{\mathcal{N}} = \begin{bmatrix} 0 & 0 \\ 0 & 1 \end{bmatrix}, \quad \text{and}$$

$$p_{\mathcal{B}} = \begin{bmatrix} (1 + A^*A)^{-1} & A^*(1 + AA^*)^{-1} \\ A(1 + A^*A)^{-1} & AA^*(1 + AA^*)^{-1} \end{bmatrix},$$

where 1 is to be interpreted as $1_{\mathcal{M}}$ of $1_{\mathcal{N}}$, as the case may warrant. Conclude that the range of $p_{\mathcal{B}}\, p_{\mathcal{M}}$ is the graph of the restriction of $A$ to dom $A^*A$, and consequently dense in $\mathcal{B}$. (Hint: if $(\xi,\eta) \in \mathcal{H}$, there exists a unique $\xi \in \text{dom } A$ and $\eta \in \text{dom } A^*$ such that $(\xi,\eta) = (\xi,A\xi) + (-A^*\eta,\eta)$; to compute $p_{\mathcal{B}}$, solve for $\xi$, in case $\eta = 0$ or $\xi = 0$; for the last assertion, you would need the fact that for $H \geqslant 0$, the graph of $H$ restricted to $\mathcal{D}_0$ is dense in the graph of $H$, where

$$\mathcal{D}_0 = \bigcup_{n=1}^{\infty} \text{ran } 1_{[0,n]}(H);$$

this last assertion follows readily from the spectral theorem -- cf. Ex. (2.5.5).)

**Lemma 1.2.7.** Let $\mathcal{M}$, $\mathcal{N}$, $\mathcal{B} \,\eta\, M$ and $\mathcal{B} \subseteq \mathcal{M} \oplus \mathcal{N}$, *with* $\mathcal{M} \perp \mathcal{N}$. *Then either* $\mathcal{B} \precsim \mathcal{M}$ *or* $(\mathcal{M} \oplus \mathcal{N}) \ominus \mathcal{B} \precsim \mathcal{N}$.

**Proof.** Let $\mathcal{M}_i$, $\mathcal{N}_i$, $\mathcal{B}_0$, $A$ be as in Lemma 1.2.5.
Case (i): $\mathcal{N}_2 \precsim \mathcal{M}_3$. Then

$$\begin{aligned} \mathcal{B} &= \mathcal{B}_0 \oplus \mathcal{M}_2 \oplus \mathcal{N}_2 \\ &\sim \mathcal{M}_1 \oplus \mathcal{M}_2 \oplus \mathcal{N}_2 \\ &\precsim \mathcal{M}_1 \oplus \mathcal{M}_2 \oplus \mathcal{M}_3 = \mathcal{M} . \end{aligned}$$

Case (ii): $\mathcal{M}_3 \precsim \mathcal{N}_2$.

Regarding $A$ as a closed densely defined operator from $\mathcal{M}_1$ to $\mathcal{N}_1$, let $A^+$ denote the closed densely defined operator from $\mathcal{N}_1$ to $\mathcal{M}_1$ which is the adjoint of $A$. Then $A^+$, viewed as an operator in $\mathcal{H}$, is clearly affiliated to $M$; further, from the general fact about the graph of the adjoint, it is clear that $(\mathcal{M}_1 \oplus \mathcal{N}_1) \ominus \mathcal{B}_0 = \{-\eta + A^+\eta : \eta \in \text{dom } A^+\}$. Arguing exactly as in the proof of Lemma 1.2.5, it may be seen that $\mathcal{N}_1 \sim ((\mathcal{M}_1 \oplus \mathcal{N}_1) \ominus \mathcal{B}_0) \sim \mathcal{M}_1$; hence,

$$\begin{aligned} (\mathcal{M} \oplus \mathcal{N}) \ominus \mathcal{B} &= ((\mathcal{M}_1 \oplus \mathcal{N}_1) \ominus \mathcal{B}_0) \oplus \mathcal{M}_3 \oplus \mathcal{N}_3 \\ &\precsim \mathcal{N}_1 \oplus \mathcal{N}_2 \oplus \mathcal{N}_3 = \mathcal{N} . \end{aligned}$$

The proof is complete, since, by Proposition 1.1.9, one of the two cases must arise. □

**Lemma 1.2.8.** *If* $\mathcal{M}, \mathcal{N}\ \eta\ M$, $\mathcal{M} \perp \mathcal{N}$, *and* $\mathcal{M}$ *and* $\mathcal{N}$ *are both finite, then* $\mathcal{M} \oplus \mathcal{N}$ *is finite.*

**Proof.** If $\mathcal{M} \oplus \mathcal{N}$ is infinite, then, by the 'chief characteristics of infinity', there exists $\mathcal{B}\ \eta\ M$ such that $\mathcal{B} \subseteq \mathcal{M} \oplus \mathcal{N}$ and $(\mathcal{M} \oplus \mathcal{N}) \sim \mathcal{B} \sim ((\mathcal{M} \oplus \mathcal{N}) \ominus \mathcal{B})$. So, by Lemma 1.2.7, either $(\mathcal{M} \oplus \mathcal{N}) \precsim \mathcal{M}$ or $(\mathcal{M} \oplus \mathcal{N}) \precsim \mathcal{N}$. The assumed infiniteness of $\mathcal{M} \oplus \mathcal{N}$ would then contradict the finiteness of $\mathcal{M}$ or $\mathcal{N}$ (cf. Prop. 1.2.2). □

**Lemma 1.2.9.** *If* $\mathcal{M}, \mathcal{N}\ \eta\ M$, *then* $([\mathcal{M} + \mathcal{N}] \ominus \mathcal{N}) \precsim \mathcal{M}$.

**Proof.**

$$\begin{aligned} [\mathcal{M} + \mathcal{N}] \ominus \mathcal{N} &= p_{\mathcal{N}^\perp}([\mathcal{M} + \mathcal{N}]) \\ &= [\{p_{\mathcal{N}^\perp} \xi : \xi \in \mathcal{M}\}] \\ &= \overline{\text{ran } p_{\mathcal{N}^\perp} p_{\mathcal{M}}} \\ &\sim \overline{\text{ran } p_{\mathcal{M}} p_{\mathcal{N}^\perp}} \qquad \text{(by Ex. (1.1.6))} \\ &\subseteq \mathcal{M} \end{aligned}$$

□

**Theorem 1.2.10.** *If* $\mathcal{M}, \mathcal{N}\ \eta\ M$, *and if* $\mathcal{M}$ *and* $\mathcal{N}$ *are finite, so is* $[\mathcal{M} + \mathcal{N}]$; *slightly more generally, the supremum of finitely many finite projections is finite.*

**Proof.** It follows from Lemma 1.2.9 that $[\mathcal{M} + \mathcal{N}] = ([\mathcal{M} + \mathcal{N}] \ominus \mathcal{N}) \oplus \mathcal{N}$ is an expression of $[\mathcal{M} + \mathcal{N}]$ as a direct sum of mutually orthogonal finite subspaces; the first assertion of the theorem follows from 1.2.8; an easy induction argument yields the second assertion. □

## 1.3. The Dimension Function

One of the basic problems in any theory is the classification, up to isomorphism (in the appropriate category), of the several objects of the category. For the classification of factors, clearly one invariant for $M$ is provided by the totally ordered set $\mathcal{P}(M)/\sim$ of equivalence classes of projections in $M$ -- the order being induced by the order $\precsim$ on $\mathcal{P}(M)$. This section will exhibit an isomorphism $\tilde{D}$ of $\mathcal{P}(M)/\sim$ onto a closed subset of $[0,\infty]$. Such a $\tilde{D}$ would induce a function $D$: $\mathcal{P}(M) \to [0,\infty]$ satisfying $D(p_{\mathcal{M}}) \leq D(p_{\mathcal{N}})$ if and only if $\mathcal{M} \precsim \mathcal{N}$. As in the foregoing sections, the symbols $\mathcal{M}$, $\mathcal{N}$, $\mathcal{B}$ and $\mathcal{R}$ will always denote closed subspaces affiliated to a factor; also, rather than writing $D(p_{\mathcal{M}})$, we shall write $D(\mathcal{M})$. The first half of this section will be devoted to proving the following result.

**Theorem 1.3.1.** *Let $M$ be a factor. There exists a function $D$: $\mathcal{P}(M) \to [0,\infty]$ such that*

(a) $\mathcal{M} \sim \mathcal{N} \Leftrightarrow D(\mathcal{M}) = D(\mathcal{N})$;
(b) $\mathcal{M} \perp \mathcal{N} \Rightarrow D(\mathcal{M} \oplus \mathcal{N}) = D(\mathcal{M}) + D(\mathcal{N})$; *and*
(c) $\mathcal{M}$ *is finite* $\Leftrightarrow D(\mathcal{M}) < \infty$.

*Further, such a function is uniquely determined up to a positive constant multiple.* □

The following exercise will establish the validity of the theorem in a simple special case, and will justify calling $D$ a dimension function for $M$.

### Exercises

**(1.3.2).** Let $M = \mathcal{L}(\mathcal{H})$. Prove:

(a) $\mathcal{M} \sim \mathcal{N} \Leftrightarrow \dim \mathcal{M} = \dim \mathcal{N}$;
(b) $\mathcal{M}$ is finite (rel $M$) $\Leftrightarrow \dim \mathcal{M} < \infty$.
(c) the equation $D(\mathcal{M}) = \dim \mathcal{M}$ satisfies the conditions (a) - (c) of Theorem 1.3.1;
(d) if $D'$: $\mathcal{P}(\mathcal{L}(\mathcal{H})) \to [0,\infty]$ is a function satisfying (a) - (c) of Theorem 1.3.1, then $D' = cD$ where $c = D'(\mathcal{N})$, for any one-dimensional subspace $\mathcal{N}$ of $\mathcal{H}$. □

In the light of this exercise, one obvious way to attempt a proof of

Theorem 1.3.1 would lead one to seek the abstract analogue of a one-dimensional subspace. One such abstraction is afforded by the following definition.

**Definition 1.3.3.** An $\mathcal{M}\ \eta\ M$ is said to be minimal if $\mathcal{M} \neq (0)$ and if $\mathcal{N}\ \eta\ M$, $\mathcal{N} \subseteq \mathcal{M}$ imply $\mathcal{N} = (0)$ or $\mathcal{N} = \mathcal{M}$. □

It is clear that minimal projections are finite and non-zero; the trouble is that such projections may not even exist in *M*. The next definition yields a partition of the class of factors into three subclasses, depending on the availability or otherwise of certain kinds of projections in *M*.

**Definition 1.3.4..** A factor *M* is said to be of type I, II or III according as it satisfies the corresponding condition below:

(I) *M* contains a minimal projection;
(II) *M* contains no minimal projection, but does contain non-zero finite projections;
(III) *M* contains no finite non-zero projection.

It is clear from the definition that any factor is of exactly one type. We shall prove Theorem 1.3.1 by treating, in order, the types III, I and II. Before doing that, however, it will help to examine the quantitative aspects of the Euclidean algorithm established earlier (cf. Prop. 1.2.3).

**Proposition 1.3.5.** *Let* $\mathcal{M}, \mathcal{N}\ \eta\ M$; *suppose* $\mathcal{N} \neq (0)$ *and* $\mathcal{M}$ *is finite. If* $\mathcal{M} = (\oplus_{i \in I} \mathcal{N}_i) \oplus \mathcal{R}$ *with* $\mathcal{N}_i \sim \mathcal{N}$ *for all* $i \in I$ *and* $\mathcal{R} \prec \mathcal{N}$ *(as in Prop. 1.2.3), the index set I is finite and its cardinality is independent of the particular decomposition chosen.*

**Proof.** Suppose $\mathcal{M} = (\oplus_{j \in J} \mathcal{N}'_j) \oplus \mathcal{R}'$ is another such decomposition and suppose, if possible, that there exists a map $\tau\colon I \to J$ which is injective but not surjective. Let $j_0 \in J \setminus \tau(I)$; note that $\mathcal{R} \prec \mathcal{N} \sim \mathcal{N}'_{j_0}$. So, there exists $\mathcal{R}_0 \subsetneq \mathcal{N}'_{j_0}$ such that $\mathcal{R} \sim \mathcal{R}_0$. Then,

$$\begin{aligned}
\mathcal{M} &= \left[\oplus_{i \in I} \mathcal{N}_i\right] \oplus \mathcal{R} \\
&\sim \left[\oplus_{i \in I} \mathcal{N}'_{\tau(i)}\right] \oplus \mathcal{R}_0 \\
&\subsetneq \oplus_{j \in J} \mathcal{N}'_j \subseteq \mathcal{M},
\end{aligned}$$

contradicting the finiteness of $\mathcal{M}$. Both assertions follow from the non-existence of a $\tau$ as above for any pair of admissible decompositions. □

**Definition 1.3.6.** If $\mathcal{M}$, $\mathcal{N}$ $\eta$ $M$ are both non-zero and finite, let $[\mathcal{M}/\mathcal{N}]$ denote the uniquely determined integer card $I$, as in Prop. 1.3.5. □

Note that in the example $M = \mathfrak{L}(\mathcal{H})$, $[\mathcal{M}/\mathcal{N}]$ is the greatest integer which does not exceed dim $\mathcal{M}$/dim $\mathcal{N}$, so the similarity with the notation for the greatest integer function ($[t] = n$ iff $n \leqslant t < n + 1$) is not an accident. Let us now proceed to the proof of Theorem 1.3.1.

**Type III.** For existence, define

$$D(\mathcal{M}) = \begin{cases} 0\,, & \text{if } \mathcal{M} = (0) \\ \infty\,, & \text{if } \mathcal{M} \neq (0). \end{cases}$$

Since, by hypothesis, every non-zero $\mathcal{M}$ $\eta$ $M$ is infinite, it is clear that $D$ satisfies the conditions (a) - (c) of Theorem 1.3.1. Conversely, if $D'$ is any function satisfying (a) - (c), then $D'(\{0\}) < \infty$ by (c) and $D'(\{0\}) = 2D'(\{0\})$ by (b), so that $D'(\{0\}) = 0$; since $M$ is of type III, (c) ensures that $D'(\mathcal{M}) = \infty$ if $\mathcal{M} \neq \{0\}$ and thus $D' = D$.

**Type I.** Let $\mathcal{N}$ be minimal, and define

$$D_{\mathcal{N}}(\mathcal{M}) = \begin{cases} \infty\,, & \text{if } \mathcal{M} \text{ is infinite} \\ [\mathcal{M}/\mathcal{N}]\,, & \text{if } \mathcal{M} \text{ is finite.} \end{cases}$$

For any $\mathcal{M}$ $\eta$ $M$, let $\mathcal{M} = (\oplus_{i\in I} \mathcal{N}_i) \oplus \mathcal{R}$ be a decomposition of $\mathcal{M}$ such that $\mathcal{R} \prec \mathcal{N} \sim \mathcal{N}_i$ for all $i$. Since $\mathcal{N}$ is minimal, conclude that $\mathcal{R} = 0$. The set $I$ is countable since $\mathcal{H}$ is separable; if $I$ is infinite, say $I = \{1,2, ..., \}$, then

$$\mathcal{M} = \overset{\infty}{\underset{n=1}{\oplus}} \mathcal{N}_n \sim \overset{\infty}{\underset{n=2}{\oplus}} \mathcal{N}_n \subsetneqq \mathcal{M}\,,$$

whence $\mathcal{M}$ is infinite; on the other hand, if $I$ is finite, then $\mathcal{M}$ is finite, by Theorem 1.2.10. Hence, in either case $D_{\mathcal{N}}(\mathcal{M})$ = card $I$. It is easy, now, to verify that (a) $D_{\mathcal{N}}$ satisfies conditions (a) - (c) of Theorem 1.3.1, and (b) if $D$ is any function satisfying (a) - (c) of Theorem 1.3.1, then $D(\mathcal{M}) = [\mathcal{M}/\mathcal{N}]D(\mathcal{N})$ for every finite $\mathcal{M}$, and consequently, $D = D(\mathcal{N})D_{\mathcal{N}}$.

**Type II.** the construction in this case is a little more involved. We begin with a lemma.

**Lemma 1.3.7.** *If $M$ is a factor of type* II, *there exists a sequence $\{\mathcal{N}_n\}_{n=1}^{\infty}$ of finite non-zero subspaces ($\eta M$) such that $[\mathcal{N}_n/\mathcal{N}_{n+1}] \geqslant 2$ for all $n$.*

**Proof.** It suffices to prove that if $\mathcal{N}$ is a non-zero finite subspace ($\eta M$), then there exists a finite non-zero $\mathcal{N}'$ such that $[\mathcal{N}/\mathcal{N}'] \geqslant 2$;

then the $\mathcal{N}_n$'s can be inductively defined. Since $\mathcal{N}$ is not minimal ($M$ being of type II), there exists $\mathcal{B}\,\eta M$ such that $(0) \neq \mathcal{B} \subsetneq \mathcal{N}$; the finiteness of $\mathcal{N}$ ensures finiteness of $\mathcal{B}$. If $[\mathcal{N}/\mathcal{B}] \geq 2$, set $\mathcal{N}' = \mathcal{B}$; if $[\mathcal{N}/\mathcal{B}] = 1$ -- note that $[\mathcal{N}/\mathcal{B}] > 0$ -- then $\mathcal{N} = \mathcal{B} \oplus \mathcal{R}$ with $\mathcal{R} \precsim \mathcal{B}$; further $\mathcal{R} \neq (0)$ since $\mathcal{B} \neq \mathcal{N}$; note that $[\mathcal{N}/\mathcal{R}] \geq 2$ and set $\mathcal{N}' = \mathcal{R}$. □

**Definition 1.3.8.** A sequence $\mathcal{S} = \{\mathcal{N}_n\}_{n=1}^{\infty}$ as in Lemma 1.3.7 will be called a fundamental sequence for the type II factor $M$. □

The following bit of notation will facilitate some of the subsequent proofs: let us agree to write $k\mathcal{N}$ for any subspace of the form $\mathcal{N}_1 \oplus \dots \oplus \mathcal{N}_k$, with $\mathcal{N}_i \sim \mathcal{N}$ for all $i$. Thus, for example, if $\mathcal{M}$ and $\mathcal{N}$ are finite and non-zero, then

$$\left[\frac{\mathcal{M}}{\mathcal{N}}\right]\mathcal{N} \precsim \mathcal{M} \precsim \left(\left[\frac{\mathcal{M}}{\mathcal{N}}\right] + 1\right)\mathcal{N}.$$

**Lemma 1.3.9.** *Let $\mathcal{M}$, $\mathcal{N}$, $\mathcal{B}$ be finite and non-zero.*

(a) $$\left[\frac{\mathcal{B}}{\mathcal{M}}\right]\left[\frac{\mathcal{M}}{\mathcal{N}}\right] \leq \left[\frac{\mathcal{B}}{\mathcal{N}}\right] < \left(\left[\frac{\mathcal{B}}{\mathcal{M}}\right] + 1\right)\left(\left[\frac{\mathcal{M}}{\mathcal{N}}\right] + 1\right);$$

(b) *if $\mathcal{M} \perp \mathcal{B}$, then*

$$\left[\frac{\mathcal{M}}{\mathcal{N}}\right] + \left[\frac{\mathcal{B}}{\mathcal{N}}\right] \leq \left[\frac{\mathcal{M} \oplus \mathcal{B}}{\mathcal{N}}\right] < \left[\frac{\mathcal{M}}{\mathcal{N}}\right] + \left[\frac{\mathcal{B}}{\mathcal{N}}\right] + 2.$$

**Proof.** Note that $[\mathcal{M}/\mathcal{N}]$ is precisely the largest number of pairwise orthogonal copies of $\mathcal{N}$ which can be fitted into $\mathcal{M}$. The first inequality, of both (a) and (b), is an immediate consequence. Turn to the second:

(a) The inequality $[\mathcal{B}/\mathcal{N}] \geq ([\mathcal{B}/\mathcal{M}] + 1)([\mathcal{M}/\mathcal{N}] + 1)$ would imply the existence, inside $\mathcal{B}$, of $([\mathcal{B}/\mathcal{M}] + 1)([\mathcal{M}/\mathcal{N}] + 1)$ pairwise orthogonal copies of $\mathcal{N}$, and consequently, if $([\mathcal{B}/\mathcal{M}] + 1)$ pairwise orthogonal copies of $\mathcal{M}$ (since $\mathcal{M} \precsim ([\mathcal{M}/\mathcal{N}] + 1)\mathcal{N}$) which is a contradiction.

(b) By the parenthetical comment in the proof of (a),

$$\mathcal{M} \oplus \mathcal{B} \precsim \left(\left[\frac{\mathcal{M}}{\mathcal{N}}\right] + 1 + \left[\frac{\mathcal{B}}{\mathcal{N}}\right] + 1\right)\mathcal{N}.$$

(Strictly speaking, this is valid only if there exist $([\mathcal{M}/\mathcal{N}] + [\mathcal{B}/\mathcal{N}] + 2)$ pairwise orthogonal copies of $\mathcal{N}$ in $\mathcal{H}$; if that is not true, then $\mathcal{H}$ must be finite, with

$$\left[\frac{\mathcal{H}}{\mathcal{N}}\right] < \left[\frac{\mathcal{M}}{\mathcal{N}}\right] + \left[\frac{\mathcal{B}}{\mathcal{N}}\right] + 2,$$

in which case the desired inequality follows since $[\mathcal{R}/\mathcal{N}]$ is clearly monotone in $\mathcal{R}$.) Since finiteness is inherited by $\mathcal{M} \oplus \mathcal{B}$ from that of $\mathcal{M}$ and $\mathcal{B}$, the desired inequality cannot be false. □

**Proposition 1.3.10.** *Let $\{\mathcal{N}_n\}_{n=1}^{\infty}$ be a fundamental sequence for $M$ and* *let* $\mathcal{M}$, $\mathcal{B}$ $\eta$ $M$ *be finite and non-zero. Then,*

(a) $\left[\frac{\mathcal{M}}{\mathcal{N}_n}\right] \neq 0$ *eventually; in fact* $\left[\frac{\mathcal{M}}{\mathcal{N}_n}\right] \nearrow +\infty$ ; *and*

(b) $\lim_{n\to\infty} \frac{[\mathcal{M}/\mathcal{N}_n]}{[\mathcal{B}/\mathcal{N}_n]}$ *exists and is a finite positive number.*

**Proof.** By Lemma 1.3.9 (a), for any $n \geqslant 1$, we have

$$\left[\frac{\mathcal{N}_1}{\mathcal{M}}\right] \geqslant \left[\frac{\mathcal{N}_1}{\mathcal{N}_n}\right]\left[\frac{\mathcal{N}_n}{\mathcal{M}}\right] \geqslant 2^{n-1}\left[\frac{\mathcal{N}_n}{\mathcal{M}}\right];$$

since $\{\mathcal{N}_1/\mathcal{M}]$ is a fixed finite integer it follows that $[\mathcal{N}_n/\mathcal{M}] = 0$ for all sufficiently large $n$; thus, there exists an integer $n_0$ such that $\mathcal{M} \not\precsim \mathcal{N}_n$ for all $n \geqslant n_0$; so, if $n \geqslant n_0$, $\mathcal{N}_n \precsim \mathcal{M}$; in other words $[\mathcal{M}/\mathcal{N}_n] \geqslant 1$. Then, for any integer $k$,

$$\left[\frac{\mathcal{M}}{\mathcal{N}_{n_0+k}}\right] \geqslant \left[\frac{\mathcal{M}}{\mathcal{N}_{n_0}}\right]\left[\frac{\mathcal{N}_{n_0}}{\mathcal{N}_{n_0+k}}\right] \geqslant 2^{k-1},$$

and (a) is proved.

(b) Appealing to both the inequalities in Lemma 1.3.9 (a), we see that if $n$ is so large that $[\mathcal{B}/\mathcal{N}_n] \geqslant 1$,

$$(*) \qquad \frac{[\mathcal{M}/\mathcal{N}_{n+k}]}{[\mathcal{B}/\mathcal{N}_{n+k}]} \leqslant \left\{\frac{[\mathcal{M}/\mathcal{N}_n] + 1}{[\mathcal{B}/\mathcal{N}_n]}\right\} \cdot \left\{\frac{[\mathcal{N}_n/\mathcal{N}_{n+k}] + 1}{[\mathcal{N}_n/\mathcal{N}_{n+k}]}\right\}$$

for any integer $k \geqslant 1$. Writing $\alpha_n = [\mathcal{M}/\mathcal{N}_n]/[\mathcal{B}/\mathcal{N}_n]$, it follows from part (a) that $0 < \alpha_n < \infty$ for $n$ large enough, and from the above inequality and part (a) that

$$\limsup_{p\to\infty} \alpha_p \leqslant \alpha_n ;$$

by varying $n$, conclude that $\lim\sup \alpha_p \leqslant \lim\inf \alpha_n$. Hence $\lim_{n\to\infty}\alpha_n$ exists and is finite. By interchanging the roles of $\mathcal{M}$ and $\mathcal{B}$, it is seen that $\lim \alpha_n > 0$. □

**End of Proof of Lemma 1.3.1.**. Let $M$ be a factor of type II and $S = \{\mathcal{N}_n\}_{n=1}^{\infty}$ be a fundamental sequence for $M$. If $\mathcal{M}$ and $\mathcal{B}$ are finite and non-zero, define $(\mathcal{M}/\mathcal{B})_S$ to be the limit whose existence is guaranteed by Prop. 1.3.10 (b). If $\mathcal{M}$, $\mathcal{B}$ and $\mathcal{N}$ are finite and non-zero, the function $(\div)_S$ is easily seen to satisfy the following conditions:

(i) $\mathcal{M} \sim \mathcal{N} \Rightarrow \left(\frac{\mathcal{M}}{\mathcal{B}}\right)_S = \left(\frac{\mathcal{N}}{\mathcal{B}}\right)_S$;

(ii) $\left(\frac{\mathcal{M}}{\mathcal{M}}\right)_S = 1$; $\left(\frac{\mathcal{B}}{\mathcal{N}}\right)_S = \left(\frac{\mathcal{B}}{\mathcal{M}}\right)_S\left(\frac{\mathcal{M}}{\mathcal{N}}\right)_S$; $\left(\frac{\mathcal{M}}{\mathcal{B}}\right)_S = \left(\frac{\mathcal{B}}{\mathcal{M}}\right)_S^{-1}$;

(iii) $M \perp \mathcal{B} \Rightarrow \left[\frac{M \oplus \mathcal{B}}{\mathcal{N}}\right]_S = \left[\frac{M}{\mathcal{N}}\right]_S + \left[\frac{\mathcal{B}}{\mathcal{N}}\right]_S$; (use Lemma 1.3.9 (b))

(iv) $M \precsim \mathcal{B} \Rightarrow \left[\frac{M}{\mathcal{N}}\right]_S \leqslant \left[\frac{\mathcal{B}}{\mathcal{N}}\right]_S$.

Now fix a finite non-zero $\mathcal{B}$ and define

$$D_S(M) = \begin{cases} 0 & , \text{ if } M = (0) \\ (M/\mathcal{B})_S, & \text{if } M \text{ is finite and non-zero} \\ \infty & , \text{ if } M \text{ is infinite.} \end{cases}$$

It is readily verified that $D_S$ satisfies the conditions (a) - (c) of Theorem 1.3.1.

Conversely, if $D$: $\mathcal{P}(M) \to [0,\infty]$ is any function satisfying (a) - (c) of Theorem 1.3.1, it is clear that for finite non-zero $M$ and $\mathcal{N}$,

$$\left[\frac{M}{\mathcal{N}}\right] D(\mathcal{N}) \leqslant D(M) \leqslant \left(\left[\frac{M}{\mathcal{N}}\right] + 1\right) D(\mathcal{N});$$

consequently, for finite non-zero $M$, and $n = 1,2, \ldots,$

$$\frac{[M/\mathcal{N}_n]}{[\mathcal{B}/\mathcal{N}_n] + 1} \leqslant \frac{D(M)}{D(\mathcal{B})} \leqslant \frac{[M/\mathcal{N}_n] + 1}{[\mathcal{B}/\mathcal{N}_n]};$$

let $n \to \infty$, recall that $[\mathcal{B}/\mathcal{N}_n] \to \infty$ and conclude that $D(M) = D(\mathcal{B})D_S(M)$. Since $D(M) = D_S(M) = \infty$ for infinite $M$, conclude that $D = D(\mathcal{B})D_S$. □

**Proposition 1.3.11.** *Let $M$ be a factor and $D$: $\mathcal{P}(M) \to [0,\infty]$ as in Theorem* 1.3.1. *Then,*

(a) $M \precsim \mathcal{N} \Leftrightarrow D(M) \leqslant D(\mathcal{N})$; *and*
(b) *$D$ is countably additive — i.e., if $\{M_n\}$ is a sequence of pairwise orthogonal subspaces ($\eta M$) and if $M = \oplus M_n$, then $D(M) = \Sigma D(M_n)$.*

**Proof.** The conditions (a) and (b) of Theorem 1.3.1 imply that $M \precsim \mathcal{N} \Rightarrow D(M) \leqslant D(\mathcal{N})$; coupled with (a), this yields: $M < \mathcal{N} \Rightarrow D(M) < D(\mathcal{N})$; thus, $M \precsim \mathcal{N}$ (resp., $M \succ \mathcal{N}$) implies that $D(M) \leqslant D(\mathcal{N})$ (resp., $D(M) > D(\mathcal{N})$). Since the possibilities $M \prec \mathcal{N}$, $M \sim \mathcal{N}$ and $M \succ \mathcal{N}$ are mutually exclusive and exhaustive, as are the possibilities $D(M) < D(\mathcal{N})$, $D(M) = D(\mathcal{N})$ and $D(M) > D(\mathcal{N})$, the reverse implication in (a) follows.

For finite sequences, the assertion (b) is a consequence of the assumed finite additivity (cf. (b) of Theorem 1.3.1) of $D$. Assume, then, that the sequence $\{M_n\}$ is infinite and that $M_n \neq (0)$ for all $n$. Finite additivity and monotonicity of $D$ show that, for all $N$,

$$\sum_{n=1}^{N} D(M_n) = D\left(\bigoplus_{n=1}^{N} M_n\right) \leqslant D(M);$$

consequently $\Sigma D(M_n) \leqslant D(M)$.

If possible, let $\Sigma D(M_n) < D(M)$. Then $\Sigma D(M_n) < \infty$; in particular, for each $\epsilon > 0$, there exists a finite non-zero $\mathcal{N} \,\eta\, M$ such that $D(\mathcal{N}) < \epsilon$.

Pick such an $\mathcal{N}$ for a fixed $\epsilon < D(\mathcal{M}) - \Sigma D(\mathcal{M}_n)$.

Since $D$ is finitely additive, note that

$$D(\mathcal{M}) - \sum_{n=1}^{\infty} D(\mathcal{M}_n) = D\left[ \bigoplus_{n=N}^{\infty} \mathcal{M}_n \right] - \sum_{n=N}^{\infty} D(\mathcal{M}_n)$$

for each $N$. So, we may assume without loss of generality -- by replacing $\{\mathcal{M}_n\}$ by $\{\mathcal{M}_{n+N}\}$, for large $N$ -- that $\Sigma D(\mathcal{M}_n) < D(\mathcal{N})$.

**Assertion**: There exists a sequence $\{\mathcal{M}_n'\}$ of pairwise orthogonal subspaces of $\mathcal{N}$ such that $\mathcal{M}_n \sim \mathcal{M}_n'$ for all $n$.

We shall construct the $\mathcal{M}_n'$ inductively. To start with, $D(\mathcal{M}_1) < D(\mathcal{N})$ implies $\mathcal{M}_1 \nsucc \mathcal{N}$ and so there exists $\mathcal{M}_1' \,\eta\, M$ such that $\mathcal{M}_1 \sim \mathcal{M}_1' \subseteq \mathcal{N}$. If, now, $\mathcal{M}_1', \ldots, \mathcal{M}_n'$ have been chosen satisfying $\mathcal{M}_i' \perp \mathcal{M}_j'$ for $1 \leqslant i < j \leqslant n$, and $\mathcal{M}_i \sim \mathcal{M}_i' \subseteq \mathcal{N}$ for $1 \leqslant i \leqslant n$, then,

$$\begin{aligned} D\left\{\mathcal{N} \ominus \left[ \bigoplus_{j=1}^{n} \mathcal{M}_j' \right] \right\} &= D(\mathcal{N}) - \sum_{j=1}^{n} D(\mathcal{M}_j') \\ &> \sum_{j>n} D(\mathcal{M}_j) \geqslant D(\mathcal{M}_{n+1}); \end{aligned}$$

so there exists $\mathcal{M}_{n+1}' \,\eta\, M$ such that

$$\mathcal{M}_{n+1} \sim \mathcal{M}_{n+1}' \subseteq \mathcal{N} \ominus \left[ \bigoplus_{j=1}^{n} \mathcal{M}_j' \right],$$

and the assertion is verified.

Since $\sim$ is countably additive, conclude that $\mathcal{M} \sim \oplus \mathcal{M}_n' \subseteq \mathcal{N}$, and hence $D(\mathcal{M}) \leqslant D(\mathcal{N}) < \epsilon = D(\mathcal{M}) - \Sigma D(\mathcal{M}_n) \leqslant D(\mathcal{M})$. This contradiction completes the proof. □

**Definition 1.3.12.** Any function $D$ as in Theorem 1.3.1 -- there are not too many of them! -- is called a dimension function of $M$. (Murray and von Neumann call it a "relative dimension function"; we dispense with the adjective "relative", one justification for such impertinence being: who has ever heard of a relative Haar measure?) □

Let us continue the analysis a little further by considering the possibilities that are open for the set $\Delta = \{D(\mathcal{M}): \mathcal{M} \,\eta\, M\}$.

**Lemma 1.3.13.** *Let $D$, $\Delta$ be as above and let $\bar{\alpha} = D(\mathcal{H})$. Then,*

(a) $\Delta \subseteq [0,\bar{\alpha}]$;
(b) $\alpha,\beta \in \Delta$ *and* $\beta < \alpha \Rightarrow \alpha - \beta \in \Delta$; *and*
(c) $\alpha_1,\alpha_2, \ldots \in \Delta$ *and* $\Sigma\alpha_n \leqslant \bar{\alpha} \Rightarrow \Sigma\alpha_n \in \Delta$.

**Proof.** Exercise! □

**Proposition 1.3.14.** *Let $D$,$\Delta$ be as above. Then $\Delta$ is one and only one of the following sets:*

($I_n$) $\{0, \bar{\epsilon}, 2\bar{\epsilon}, ..., n\bar{\epsilon}\}$, *where* $0 < \bar{\epsilon} < \infty$; $(n = 1,2, ...)$
($I_\infty$) $\{n\bar{\epsilon} : n = 0,1,2, ..., \infty\}$, *where* $0 < \bar{\epsilon} < \infty$
($II_1$) $[0,\bar{\alpha}]$, *where* $0 < \bar{\alpha} < \infty$
($II_\infty$) $[0, \infty]$
(III) $\{0, \infty\}$.

**Proof.** We consider, separately, the three cases corresponding to the possible type of $M$.

**Case (i):** $M$ is of type I.
Let $N \;\eta\; M$ be minimal. We have already seen (in the proof of Theorem 1.3.1 for type I) that any $\mathcal{M} \;\eta\; M$ is of the form $\mathcal{M} = \oplus_{i \in I} N_i$, with $I$ countable and $N_i \sim N$ for each $i$, and so $\Delta \subseteq \{n\bar{\epsilon} : n = 0,1,2, ...\} \cup \{\infty\}$ where $\bar{\epsilon} = D(N)$. It is easy to see that (i) if $\mathcal{H}$ is finite and $[\mathcal{H}/N] = n$, then $\Delta = \{k\bar{\epsilon} : k = 0,1, ..., n\}$ and (ii) if $\mathcal{H}$ is infinite, then $\Delta = \{n\bar{\epsilon} : n = 0,1, ..., \infty\}$.

**Case (ii):** $M$ is of type II.
Since $M$ has a fundamental sequence, it follows that $\Delta$ does not contain a smallest positive number. Let $\bar{\alpha} = D(\mathcal{H})$. Infer from Lemma 1.3.13 that if $\alpha \in \Delta$, then $k\alpha \in \Delta$ for any integer $k$ such that $k\alpha \leqslant \bar{\alpha}$. Taken together, the preceding two sentences guarantee that $\Delta$ is dense in $[0,\bar{\alpha}]$. If now, $0 < \alpha < \bar{\alpha}$, pick a sequence $\{\alpha_n\} \subseteq \Delta$ such that $\alpha_n \nearrow \alpha$. So, by Lemma 1.3.13, $\alpha = \Sigma(\alpha_n - \alpha_{n-1})$ (with $\alpha_0 = 0$) $\in \Delta$. This proves $\Delta \supseteq [0,\bar{\alpha}]$, the other inclusion being trivial.

**Case (iii):** $M$ is of type III.
Clearly, in this case $\Delta = \{0,\infty\}$. □

**Definition 1.3.15.** A factor $M$ is said to be of type $I_n$, $I_\infty$, $II_1$, $II_\infty$ or III according as the range of the dimension function of $M$ satisfies the corresponding condition of Proposition 1.3.14. Factors of type $I_n$ $(n < \infty)$ and $II_1$ are called finite, and the other types are called infinite. (Thus, a factor $M$ is finite if and only if $\mathcal{H}$ is finite (rel $M$).) Factors of type I or II are said to be semifinite. (Thus a factor $M$ is semifinite if and only if it contains non-zero finite projections.) □

Examples will be given later in Section 4.3 to show that factors of all these types exist.

No treatment of the Murray-von Neumann papers would be complete without at least a passing mention of the so-called reduction theory, whereby every von Neumann algebra is expressed as a direct integral of factors. Very briefly, one uses the abelian von Neumann algebra $Z(M)$ to represent the underlying Hilbert space as a direct integral of Hilbert spaces over a measure space in such a way that $Z(M)$ acts as 'scalar decomposable' operators. (Actually, the underlying Borel space may be taken to be (a compact subset of) the real line, since every abelian von Neumann algebra acting on a separable Hilbert space is generated by a single self-adjoint operator -- but that is not really crucial.)

The theory goes on to show that if $\mathcal{H} = \int^{\oplus}\mathcal{H}(\lambda)d\mu(\lambda)$, there is, for each $\lambda$, a factor $M(\lambda) \subseteq \mathfrak{L}(\mathcal{H}(\lambda))$, the assignment $\lambda \to M(\lambda)$ being "measurable" in a certain sense, so that $M$ is the collection of operators of the form $x = \int^{\oplus}x(\lambda)d\mu(\lambda)$, where $x(\lambda) \in M(\lambda)$, the map $x(\cdot)$ being measurable in an appropriate (weak) sense and satisfying $\|x\| = \text{ess. sup}\|x(\cdot)\| < \infty$. Using this theory, one may speak of the type of a general von Neumann algebra; call $M$ type $\text{II}_1$, for instance, if each (i.e., a.e.) $M(\lambda)$ is of type $\text{II}_1$ and so on.

After deliberating on whether or not to devote a section in this chapter to a more elaborate exposition of this theory, the author opted for "not to," on the following counts: (a) the material is not really pertinent to the remainder of the book; (b) it is not really necessary to torment the uninitiated reader with the spectre of non-measurability that is inescapable in anything like a serious discussion of disintegration; and (c) the initiated reader does not need the section anyway.

The interested reader should go directly to the fountainhead for as readable and self-contained an exposition of the theory as is possible.

# Chapter 2
# THE TOMITA-TAKESAKI THEORY

Section 2.1 discusses the following question (which, in the case $M = L^\infty(X,\mathcal{F},\mu)$ with $\mu$ finite, is answered affirmatively by the existence of the Lebesgue integral): if $m$: $\mathcal{P}(M) \to [0,1]$ is countably additive in the sense that

$$m\left(\bigvee_{n=1}^{\infty} e_n\right) = \sum_{n=1}^{\infty} m(e_n)$$

for any countable collection of pairwise orthogonal projections in $M$, does $m$ extend to a linear functional on $M$ which is well-behaved under monotone convergence?

Section 2.2 is devoted to the celebrated GNS construction, which, in case $M = L^\infty(X,\mathcal{F},\mu)$ yields the Hilbert space $L^2(X,\mathcal{F},\mu)$ and the representation of $M$ as multiplication operators.

Section 2.3 is concerned with the (conjugate linear) operator on the GNS space which is induced by the map $x \to x^*$ of $M$. The climax is the Tomita-Takesaki theorem which involves a thorough analysis of the antiunitary and positive factors in the polar decomposition of the above mentioned operator, and their commutation relations with the operators in $M$. One crucial fact emerging from this theorem is the existence of a certain one-parameter group (called the modular group) of automorphisms of $M$, which is of fundamental importance when $M$ is of type III. As the proof of the theorem is long and technical, the proof is presented only in the very special case when the above-mentioned operator is bounded. Although this case never arises when $M$ is of type III (as will be established later), this option has been taken as a compromise between no proof and complete proof, both alternatives being distasteful to the author.

Section 2.4 introduces weights, which are non-commutative analogues of infinite measures. There are few proofs in this section. The results are stated and some tentative effects are made at convincing the reader that surely the stated results are plausible enough. The statement of the Tomita-Takesaki theorem in its full generality also makes an appearance in this section.

The next section pertains to a very useful technical criterion, called the KMS boundary condition, which gives an intrinsic characterisation (that does not appeal to the GNS construction) of the modular group associated with a weight.

The chapter ends with a discussion of (a) the noncommutative Radon-Nikodym theorem of Pedersen and Takesaki, and (b) conditional expectations and Takesaki's theorem which identifies those situations in which normal conditional expectations exist.

## 2.1. Noncommutative Integration

The symbol $M$ will always denote a von Neumann algebra of operators on a separable Hilbert space. The collection of positive operators in $M$ will be denoted by $M_+$. The positive cone $M_+$ defines an order on the real vector space $M_h$ of self-adjoint elements of $M$, whereby $x \leqslant y$ precisely when $y - x \in M_+$. Thus, for instance, $x \in M_+$ and $x \geqslant 0$ are equivalent, as of course they should be.

A linear functional $\phi$ on $M$ is said to be positive if $\phi(x^*x) \geqslant 0$ for all $x$ in $M$, or, equivalently, if $\phi(M_+) \subseteq \mathbb{R}_+$. The collection of such $\phi$ will be denoted by $M_+^*$. An element $\phi$ of $M_+^*$ is called a state if it is normalized so that $\phi(1) = 1$.

### Exercises

**(2.1.1)** Let $\phi \in M_+^*$.

(a) The equation $[x,y] = \phi(y^*x)$ defines a sesquilinear positive semi-definite form on $M$.

(b) $\phi$ satisfies the Cauchy-Schwarz inequality:

$$|\phi(y^*x)| \leqslant \phi(x^*x)^{1/2}\phi(y^*y)^{1/2}.$$

(c) $\phi$ is bounded and $\|\phi\| = \phi(1)$. (Hint: put $y = 1$ in (b) and use $x^*x \leqslant \|x\|^2\, 1$.)

(It is true, conversely, that any bounded linear function on $M$, which attains its norm at the identity, is automatically positive; a proof of this may be found in [Arv 1], for instance.) □

**Definition 2.1.2.** A positive linear functional $\phi$ on $M$ is said to be

(i) faithful if $0 \neq x \in M_+$ implies $\phi(x) > 0$;
(ii) normal if $\phi(x) = \sup_i \phi(x_i)$, whenever $x$ is the supremum of a monotone increasing net $\{x_i\}$ in $M_+$ (cf. Prop. 0.4.11);
(iii) tracial if $\phi(x^*x) = \phi(xx^*)$ for all $x$ in $M$. □

**Exercises**

**(2.1.3)** If $\phi \in M_+^*$, show that the following conditions are equivalent:

(i) $\phi$ is tracial;
(ii) $\phi(xy) = \phi(yx)$ for all $x,y$ in $M$;
(iii) $\phi(uxu^*) = \phi(x)$ for all $x \in M$ and unitary $u \in M$.

(Hint: Use polarization for (i) $\Rightarrow$ (ii) and prove (iii) $\Rightarrow$ (ii) by re-stating (iii) as $\phi(ux) = \phi(xu)$ and using the fact that the unitary operators in $M$ span $M$ as a vector space.)

**(2.1.4)** Let $M = L^\infty(X,\mathcal{F},\mu)$, and let $\phi \in M_+^*$. (We have identified $\psi$ with $m_\psi$, in the notation of Ex. (0.4.5).)

(a) The equation $\nu(E) = \phi(1_E)$ defines a finitely additive measure on $(X,\mathcal{F})$, which is absolutely continuous with respect to $\mu$.
(b) $\phi$ is normal if and only if $\nu$ (as above) is countably additive, in which case $\phi(f) = \int fg \, d\mu$ for some non-negative $g \in L^1(X,\mu)$;
(c) $\phi$ is faithful if and only if $\mu$ is absolutely continuous with respect to $\nu$. □

Exercise (2.1.4) (b) is a special case of a more general fact: a positive linear functional on $M$ is normal if and only if it is $\sigma$-weakly continuous. We shall omit a proof of this fact -- one may be found in [Dix], for instance -- but will freely use it in the sequel. The collection of normal positive linear functionals will be denoted by $M_{*,+}$. It is not too hard to establish that $M_+$ and $M_{*,+}$ are dual cones -- i.e., if $x \in M$, then $x \in M_+$ if and only if $\phi(x) \geq 0$ for all $\phi$ in $M_{*,+}$, and similarly, the dual statement (with the roles of $\phi$ and $x$ interchanged) is also valid. (As above, we shall think of the elements of $M_*$ as linear functionals on $M$.)

**Exercises**

**(2.1.5)** Let $\phi \in M_{*,+}$, let $x \in M$ and let $M_0 = \{x\}''$, the von Neumann algebra generated by $x$ and 1.

(a) If $x$ is normal, the equation $\nu_x(E) = \phi(1_E(x))$ defines a measure on the spectrum of $x$ such that $\int f \, d\nu_x = \phi(f(x))$ for every bounded Borel function $f$ defined on sp $x$.
(b) For general $x$, show that if $\phi$ is tracial, then, with the notation of (a), $\nu_{|x|} = \nu_{|x^*|}$. □

The reader will notice much later, in Section 4.3 that the verification that a given finite factor is finite will invariably be accomplished by first constructing a faithful normal tracial positive linear functional $\tau$ on $M$, and then observing that the restriction of $\tau$ to $P(M)$ yields a dimension function for $M$ which assigns a finite value to 1.

Consider now the converse problem as to whether the dimension function of a finite factor extends to a normal positive linear functional on $M$, which, if it exists, would automatically be faithful and tracial. The preceding exercises suggest that if $x \in M_h$, one could define the measure $\nu_x$ on sp $x$ by $\nu_x(E) = D(1_E(x))$, then set $\tau(x) = \int \lambda d\nu_x(\lambda)$ and attempt to establish that $\tau$ defines a real linear functional on $M_h$, and extends to a positive normal linear functional on $M$. The problem suggested at the start of the paragraph does indeed have an affirmative answer, and the resulting $\tau$ must be defined in the manner suggested above. However, the solution is by no means immediate or transparent. The following two indications should suffice to establish the non-triviality of this problem.

(a) The corresponding problem for a semifinite factor seems to have at least temporarily foxed even von Neumann, as evidenced by the fact that this is stated as Problem 11 in "On rings of operators", and settled only a year later in "Rings of operators II"; and (b) the problem of extending a general countably additive function $m$: $\mathcal{P}(M) \to [0,1]$ (not necessarily satisfying $m(u^*u) = m(uu^*)$ for partial isometries $u$) to a normal positive linear functional on $M$, in the very special case $M = \mathfrak{L}(\mathcal{H})$, is affirmatively settled by a celebrated (and definitely non-trivial) theorem due to Gleason -- even here, all the intricacies of the problem are present when dim $\mathcal{H} = 3$.

So, we shall choose to work with the integral rather than with the measure; in other words, we shall henceforth work with (normal) positive linear functionals, rather than with countably additive "measures" $m$: $\mathcal{P}(M) \to [0,1]$.

## 2.2. The GNS Construction

When $M = L^\infty(X,\mathcal{F},\mu)$, choosing a faithful normal positive linear functional on $M$ amounts to choosing a finite measure $\nu$ which is equivalent, in the sense of mutual absolute continuity, to $\mu$ [cf. Ex. (2.1.4)]. Given such a $\nu$, one can immediately construct the Hilbert space $L^2(X,\mathcal{F},\nu)$ and the associated representation of $M$ as multiplication operators. That a very similar analysis may be carried out for a general (not necessarily abelian) $M$, is the content of the celebrated Gelfand-Naimark-Segal (henceforth abbreviated to GNS) construction. This construction is valid in much greater generality -- the positive functional need not be faithful; it need not even be normal; in fact, one does not even need a von Neumann algebra, just a C*-algebra or even less, being adequate for the construction to go through. Such generality is irrelevant to our needs and we shall consider only the case of a faithful normal positive linear functional on a von Neumann algebra.

**Theorem 2.2.1.** *Let $\phi \in M_{*,+}$ be faithful. Then there exists a triple $(\mathcal{H}_\phi,\pi_\phi,\Omega_\phi)$ where*

(a) *$\pi_\phi$ is a *-algebra homomorphism of $M$ into $\mathcal{L}(\mathcal{H}_\phi)$, $\mathcal{H}_\phi$ being a Hilbert space;*
(b) *$\Omega_\phi \in \mathcal{H}_\phi$ and $\mathcal{H}_\phi = \overline{\pi_\phi(M)\Omega_\phi}$ ; and*
(c) *$\phi(x) = \langle\pi_\phi(x)\Omega_\phi,\Omega_\phi\rangle$ for all $x$ in $M$.*

*Such a triple is unique in the sense that if $(\mathcal{H}',\pi',\Omega')$ is another such triple, there exists a unique unitary operator $w$: $\mathcal{H}_\phi \to \mathcal{H}'$ such that $w\Omega_\phi = \Omega'$ and $\pi'(x) = w\pi_\phi(x)w^*$ for all $x$ in $M$.*

*Further, the image $\pi_\phi(M)$ is a von Neumann algebra of operators on $\mathcal{H}_\phi$; $\pi_\phi$ is norm-preserving as well as being a σ-weak homeomorphism of $M$ onto $\pi_\phi(M)$.*

**Proof.** If $x,y \in M$, define $[x,y] = \phi(y^*x)$. The positivity and faithfulness of $\phi$ ensure that $[\cdot,\cdot]$ is a genuine (positive-definite) inner product on $M$. Let $\mathcal{H}_\phi$ denote the completion of $M$ in this inner product. Let $\eta$: $M \to \mathcal{H}_\phi$ denote the inclusion map, and let $\eta(1) = \Omega_\phi$.

If $x,y \in M$, note that

$$\|\eta(xy)\|^2 = \phi(y^*x^*xy) \leq \|x\|^2\phi(y^*y) = \|x\|^2\ \|\eta(y)\|^2,$$

where we have used $x^*x \leq \|x\|^2 1$, and so $y^*x^*xy \leq y^*(\|x\|)^2 1)y$. So, there exists a unique bounded operator $\pi_\phi(x)$ on $\mathcal{H}_\phi$ such that $\pi_\phi(x)\eta(y) = \eta(xy)$ for all $y$ in $M$. It is easily verified that $\pi_\phi$ is a *-algebra homomorphism of $M$ into $\mathcal{L}(\mathcal{H}_\phi)$. As a sample, we show that $\pi_\phi$ preserves adjoints. For this, note that $\eta(M)$ is dense in $\mathcal{H}_\phi$, and that if $x \in M$, then for all $y,z$ in $M$,

$$\begin{aligned}\langle\pi_\phi(x)\eta(y),\eta(z)\rangle &= [xy,z] = \phi(z^*xy) = \phi((x^*z)^*y)\\ &= [y,x^*z] = \langle\eta(y),\pi_\phi(x^*)\eta(z)\rangle\ ;\end{aligned}$$

conclude that $\pi_\phi(x^*) = \pi_\phi(x)^*$. The assertions (b) and (c) of the theorem follow immediately from the definition of $\Omega_\phi$ and the inner product in $\mathcal{H}_\phi$.

If $(\mathcal{H}',\pi',\Omega')$ is another such triple, note that for $x$ and $y$ in $M$,

$$\begin{aligned}\langle\pi'(x)\Omega',\pi'(y)\Omega'\rangle &= \langle\pi'(y^*x)\Omega',\Omega'\rangle\\ &= \phi(y^*x)\\ &= \langle\pi_\phi(x)\Omega_\phi,\pi_\phi(y)\Omega_\phi\rangle\ ;\end{aligned}$$

since $\overline{\pi_\phi(M)\Omega_\phi} = \mathcal{H}_\phi$ and $\overline{\pi'(M)\Omega'} = \mathcal{H}'$, deduce the existence of a (well-defined) unitary operator $w$: $\mathcal{H}_\phi \to \mathcal{H}'$ such that $w\pi_\phi(x)\Omega_\phi = \pi'(x)\Omega'$. It is fairly clear that $w \circ \pi_\phi(x) = \pi'(x) \circ w$, since the two operators agree on the dense set $\pi_\phi(M)\Omega_\phi$, both operators mapping $\pi_\phi(y)\Omega_\phi$ to $\pi'(xy)\Omega'$; also, $w\ \Omega_\phi = w\ \pi_\phi(1)\Omega_\phi = \pi'(1)\Omega' = \Omega'$, and the second part of the theorem is proved.

Finally, if $\pi_\phi(x) = 0$, then conclude from the faithfulness of $\phi$ and the equation $\phi(x^*x) = \|\pi_\phi(x)\Omega_\phi\|^2 = 0$, that $x = 0$; thus $\pi_\phi$ is injective. Further, the map $\pi_\phi$ is normal in the sense that if $\{x_i\}$ is a monotone increasing net in $M_+$, which converges weakly to $x$, then $\pi(x_i) \to \pi(x)$ weakly. (Reason: if $x_i \nearrow x$, then $y^*x_iy \nearrow y^*xy$ for all $y$ in $M$; since $\phi$ is normal, this means that

$$<\pi_\phi(x_i)\eta_\phi(y),\eta_\phi(y)> \nearrow <\pi_\phi(x)\eta_\phi(y),\eta_\phi(y)>$$

for all $y$ in $M$; since the net $\{x_i\}$ is uniformly bounded, so is the net $\{\pi(x_i)\}$; as $\overline{\eta(M)} = \mathcal{H}_\phi$, conclude that $\pi_\phi(x_i) \nearrow \pi_\phi(x)$, as asserted.)

We shall complete the proof by showing that if $\pi$: $M \to \mathcal{L}(\mathcal{H}')$ is an injective normal *-homomorphism of $M$ into $\mathcal{L}(\mathcal{H}')$, then $\pi$ is isometric, $\pi(M)$ is $\sigma$-weakly closed and $\pi$ is a $\sigma$-weak homeomorphism of $M$ onto $\pi(M)$.

To prove $\pi$ is isometric, it suffices, thanks to the C*-identity ($\|x\|^2 = \|x^*x\|$) to verify that $\|\pi(x)\| = \|x\|$ when $x = x^* \in M$. Fix such an $x$; since the norm of a self-adjoint operator is equal to its spectral radius, it is more than sufficient to show that sp $\pi(x)$ = sp $x$. For this, if $\lambda \notin$ sp $x$, note that $(x - \lambda)^{-1} \in M$, by the double commutant theorem; hence $\pi(x) - \lambda$ is invertible, with inverse $\pi((x - \lambda)^{-1})$. Thus, sp $\pi(x) \subseteq$ sp $x$. Suppose this inclusion is strict; then there exists a continuous real function $f$ on sp $x$, which vanishes on sp $\pi(x)$ but not everywhere in sp $x$. Then $f(\pi(x)) = 0$, while $f(x) \neq 0$. This contradicts injectivity of $\pi$ since $\pi(f(x)) = f(\pi(x))$. (The proof of this last fact requires only polynomial approximation and the fact that for continuous $f$, $\|f(\pi(x))\| = \sup\{|f(\lambda)|: \lambda \in \text{sp } \pi(x)\}$.)

Next, note that if $\psi \in \mathcal{L}(\mathcal{H}')_{*,+}$ (viewed as a $\sigma$-weakly continuous linear functional on $\mathcal{L}(\mathcal{H}')$), then $\psi \circ \pi$ is a normal linear functional (since both $\psi$ and $\pi$ are normal). However, for linear functionals, recall that normality and $\sigma$-weak continuity are equivalent (cf. remarks following Ex. (2.1.4)). So, if a net $\{x_i\}$ in $M$ converges $\sigma$-weakly to $x$, then $\psi(\pi(x_i)) \to \psi(\pi(x))$. Since $\psi$ was arbitrary, $\pi(x_i) \to \pi(x)$ $\sigma$-weakly; in other words, $\pi$ is $\sigma$-weakly continuous. Since ball $M = \{x \in M: \|x\| \leq 1\}$ is $\sigma$-weakly compact and since $\pi$ is isometric, infer that ball $\pi(M) = \pi$(ball $M$) is $\sigma$-weakly compact. The Eberlein-Schmulyan theorem (cf., for instance, [Yos]) states that a linear subspace of a (Banach-) dual space is weak*-closed if and only if its unit ball is weak*-closed. Conclude that $\pi(M)$ is $\sigma$-weakly closed and consequently a von Neumann algebra. (It is a fact that a self-adjoint algebra is weakly closed if and only if it is $\sigma$-weakly closed; one proof of this uses Kaplansky's density theorem, and we do not go into that here.)

Since $\pi^{-1}$: $\pi(M) \to M$ is an injective normal *-homomorphism, it follows from the above discussion that $\pi^{-1}$ is also $\sigma$-weakly continuous, and the proof is (finally!) complete. □

It may be relevant to point out here that, more generally than was established in the above proof, it is true that an injective *-homomorphism between two (possibly abstract) C*-algebras is isometric; it may be inferred from this -- by passing through an appropriate quotient algebra -- that a *-homomorphic image of a C*-algebra is norm-closed. The interested reader may consult [Arv 1] for details.

We shall, henceforth, feel free to talk of a *-homomorphism being normal -- meaning that it preserves monotone limits -- and to use the fact, emerging from the proof of the Theorem, that a *-homomorphism is $\sigma$-weakly continuous if and only if it is normal. It is true, as in the C*-case, that the image $\pi(M)$ of a von Neumann algebra under a normal *-homomorphism is $\sigma$-weakly closed and hence a von Neumann algebra. The proof of this assertion is outlined in the following exercise.

### Exercises

**(2.2.2)** Let $I$ be a $\sigma$-weakly closed two sided ideal in $M$.

(a) Let $x \in M$ have polar decomposition $x = u|x|$. Show that the following conditions are equivalent: (i) $x \in I$; (ii) $|x| \in I$; (iii) $1_{(0,\infty)}(|x|) \in I$. Conclude from (i) $\Leftrightarrow$ (ii) that $I$ is self-adjoint. (Hint: $|x| = u^*x$ and so (i) $\Leftrightarrow$ (ii); $x = x\ 1_{(0,\infty)}(|x|)$ and so (iii) $\Rightarrow$ (i); for (ii) $\Rightarrow$ (iii), pick $y_n = f_n(|x|)$, for an appropriate Borel function $f_n$, such that $|x|y_n = 1_{(1/n,\infty)}(|x|)$, and note that

$$1_{(1/n,\infty)}(|x|) \to 1_{(0,\infty)}(|x|) \quad \sigma\text{-weakly.})$$

(b) If $x \in M_+$, then $x^{1/n} \to 1_{(0,\infty)}(x)$ $\sigma$-strongly.

(c) If $e,f \in P(M) \cap I$, then $e \vee f = \sigma\text{-strong} \lim(e + f)^{1/n}$ and so $e \vee f \in I$; in particular, $P(M) \cap I$ is an upward directed net (which is non-trivial if $I \neq \{0\}$).

(d) If $\bar{e} = \vee\{f \in P(M): f \in I\}$, then $\bar{e} \in I \cap Z(M)$. (Hint: $\bar{e} \in I$, in view of the second statement in (c) and the $\sigma$-weak closure of $I$; to see that $\bar{e}$ is central, note that if $f \in P(M) \cap I$ and $u$ is a unitary element of $M$, then $ufu^* \in P(M) \cap I$, and conclude that $u\bar{e}u^* = \bar{e}$.)

(e) Show that $I = M\bar{e}$, and that conversely, if $e \in P(Z(M))$, then $Me$ is a $\sigma$-weakly closed two-sided ideal of $M$. (Hint: The second assertion as well as the inclusion $I \supseteq M\bar{e}$ are trivial; if $x \in I$, then by (a) and the definition of $\bar{e}$, $1_{(0,\infty)}(|x|) \leq \bar{e}$ and hence $x = x\bar{e}$.)

(f) If $\pi: M \to \mathcal{L}(\mathcal{H}')$ is a normal *-homomorphism, then $\pi(M)$ is $\sigma$-weakly closed and hence a von Neumann subalgebra of $\mathcal{L}(\mathcal{H}')$. (Hint: Apply (e) to write $\ker \pi = M\bar{e}$, note that $M(1 - \bar{e})$ may be viewed as a von Neumann algebra acting on $\ker \bar{e}$, that $\pi|M(1-\bar{e})$

is injective, and appeal to the already established injective case of the assertion.) □

In view of the uniqueness assertion in Theorem 2.2.1, we shall talk in the sequel, of *the* GNS triple $(\mathcal{H}_\phi,\pi_\phi,\Omega_\phi)$ associated with a faithful normal positive linear functional $\phi$. Note incidentally that $\Omega_\phi$ is a unit vector if and only if $\phi$ is a state.

**Example 2.2.3.** (a) Let $M = L^\infty(X,\mathcal{F},\mu)$; let $\nu$ be a finite measure with the same null-sets as $\mu$; then the equation $\phi(f) = \int f d\nu$ defines a faithful normal positive linear functional $\phi$ on $M$. Two possible GNS triples are given thus:

(i) $\mathcal{H}_\phi = L^2(X,\mathcal{F},\nu)$, $\pi_\phi(f)g = fg$, $\Omega_\phi \equiv 1$; and

(ii) $\mathcal{H}' = L^2(X,\mathcal{F},\mu)$, $\pi'(f)g = fg$, $\Omega' = \left[\dfrac{d\nu}{d\mu}\right]^{1/2}$.

The unitary operator, whose existence is guaranteed by Theorem 2.2.1, is defined by $wf = f(d\nu/d\mu)^{1/2}$ for $f$ in $L^2(X,\mathcal{F},\nu)$.

(b) Let $M = \mathcal{L}(\mathcal{H})$ and let $\rho$ be an injective positive trace-class operator on $\mathcal{H}$. Then $\phi(x) = \mathrm{tr}\, \rho x$ defines a faithful normal positive linear functional on $M$. Note that if

$$\rho = \sum_{n=1}^{\infty} \alpha_n\, t_{\xi_n,\xi_n}$$

(with $\alpha_n > 0$, $\Sigma\alpha_n < \infty$ and $\{\xi_n\}$ an orthonormal basis for $\mathcal{H}$), then, for any $x$ in $M$,

$$\phi(x) = \sum_{n=1}^{\infty} \alpha_n \langle x\xi_n,\xi_n\rangle .$$

It is not hard to verify that a version of the GNS triple is given by

$$\mathcal{H}_\phi = \bigoplus_{n=1}^{\infty} \mathcal{H}_n,$$

with $\mathcal{H}_n = \mathcal{H}$ for all $n$,

$$\Omega_\phi = \bigoplus_{n=1}^{\infty} (\alpha_n^{1/2}\xi_n),\ \pi_\phi(x)(\oplus\, \eta_n) = \oplus\, (x\eta_n).$$

(Check!) An isomorphic picture is obtained by setting $\mathcal{H}_\phi = \mathcal{H} \otimes \mathcal{H}$, $\Omega_\phi = \Sigma_n\alpha_n^{1/2}\xi_n \otimes \xi_n$ and $\pi_\phi(x) = x \otimes 1$.

(c) If $M \subseteq \mathcal{L}(\mathcal{H})$ and $\rho$ is as in (b), let $\phi(x) = \mathrm{tr}\, \rho x$ for $x \in M$. This shows that every von Neumann algebra acting on a separable Hilbert space admits a faithful normal state, so that Theorem 2.2.1 is not vacuous. In this case, let $\Omega_\phi$ be as in (b) and let $\mathcal{H}_\phi = [\{(x \otimes 1)\Omega_\phi: x \in M\}] \subseteq \mathcal{H} \otimes \mathcal{H}$ and let $\pi_\phi(x)$ be the restriction to $\mathcal{H}_\phi$ of $x \otimes 1$. It is immediate that this yields one version of the GNS triple. □

More often than not, when we are given a faithful normal positive linear functional $\phi$ on $M$, we shall identify $M$ with $\pi_\phi(M)$, and thus assume that $M \subseteq \mathfrak{L}(\mathcal{H})$ and that $\phi(x) = \langle x\Omega,\Omega \rangle$ for a vector $\Omega$ in $\mathcal{H}$ such that $[M\Omega] = \mathcal{H}$. The vector $\Omega$ is known to mathematical physicists as the vacuum vector or the vacuum state (in case $\|\Omega\| = 1$). The faithfulness of $\phi$ translates to this separating property of $\Omega$: if $x \in M$, then $x = 0$ if and only if $x\Omega = 0$. Thus, in the terminology of the following definition, the vector $\Omega$ is cyclic and separating for $M$.

**Definition 2.2.4.** A set $S \subseteq \mathcal{H}$ is said to be

(i) cyclic for $M$ if $[MS] = \mathcal{H}$;
(ii) separating for $M$ if for $x$ in $M$, $x = 0$ if and only if $xS = \{0\}$. □

**Exercises**

**(2.2.5)** Let $M \subseteq \mathfrak{L}(\mathcal{H})$ and let $\phi(x) = \operatorname{tr} \rho x$, where $\rho$ is a positive trace class operator, given by

$$\rho = \Sigma\, \alpha_n\, t_{\xi_n,\xi_n},$$

with $\alpha_n > 0$ and $\{\xi_n\}$ orthonormal. Show that $\phi$ is faithful as a linear functional on $M$ if and only if $\{\xi_n\}$ is separating for $M$.

**(2.2.6)** If $S \subseteq \mathcal{H}$, show that $S$ is cyclic for $M$ if and only if $S$ is separating for $M'$. (Hint: $x' \in M'$ and $x'S = \{0\} \Rightarrow x'[MS] = \{0\}$; so, if $S$ is cyclic for $M$, $S$ is separating for $M'$. Conversely, if $S$ is separating for $M'$, note that $p' = p_{[MS]} \in M'$ and that $(1 - p')S = \{0\}$, whence $p' = 1$.)

**(2.2.7)** If $\phi(x) = \langle x\Omega,\Omega \rangle$ for $x$ in $M$, then $\phi$ is tracial if and only if $\|x\Omega\| = \|x^*\Omega\|$ for all $x$ in $M$. □

Thus, given a faithful normal positive linear functional $\phi$ on $M$, the GNS construction leads to a realization of $M$ as a von Neumann algebra of operators on a Hilbert space $\mathcal{H}_\phi$, in which there is a cyclic and separating vector for $\pi_\phi(M)$; this vector is automatically a cyclic and separating vector for $\pi_\phi(M)'$.

We shall conclude this section with an important class of von Neumann algebras which come equipped with a natural cyclic and separating vector -- the so-called group-von Neumann algebras associated with countable discrete groups.

Let $G$ be a countable discrete group, whose identity element we shall denote by $\epsilon$ (the symbols $e$ and $1$ having already been irreversibly identified with projections and the identity operator). Let $\ell^2(G)$ denote the Hilbert space of square-summable functions on $G$:

$$\ell^2(G) = \left\{ \xi: G \to \mathbb{C}: \sum_{t \in G} |\xi(t)|^2 < \infty \right\}.$$

There is a canonical orthonormal basis $\{\xi_t: t \in G\}$ of $\ell^2(G)$, where

$$\xi_t(s) = \delta_{ts} = \begin{cases} 1, & \text{if } t = s \\ 0, & \text{if } t \neq s \end{cases}.$$

For each $t \in G$, let $\lambda_t$ denote the unitary operator corresponding to left translation by $t$; thus $(\lambda_t\xi)(s) = \xi(t^{-1}s)$, or, equivalently, $\lambda_t\xi_s = \xi_{ts}$ for all $s$ in $G$. The map $t \to \lambda_t$ is a unitary representation of $G$ (i.e., $\lambda_{st} = \lambda_s\lambda_t$) in $\ell^2(G)$; it is the so-called left regular representation of $G$.

The von Neumann algebra $M = \{\lambda_t\colon t \in G\}''$ is called the group von Neumann algebra of $G$ and will be denoted by $W^*(G)$. The set $M_0$ of finite linear combinations of the $\lambda_t$'s is a self-adjoint algebra (since $\lambda_t^* = \lambda_{t^{-1}}$) containing $1(= \lambda_\epsilon)$; hence, $M$ is the strong closure of $M_0$.

(It is, in fact, true that every element of $M$ is uniquely expressible as the sum of a $\sigma$-strongly* convergent series $x = \Sigma_{t \in G} x(t)\lambda_t$, where $x$: $G \to \mathbb{C}$ satisfies $\Sigma_{t \in G}|x(t)|^2 < \infty$; this will follow from a more general assertion established later, in Section 4.1.)

Let $\Omega = \xi_\epsilon$ and observe that $\lambda_t\Omega = \xi_t$ for each $t$, and, consequently, that $\Omega$ is cyclic for $M$. We shall verify that $\Omega$ is separating for $M$, by proving that $\Omega$ is cyclic for $M'$ [cf. Ex. (2.2.6)]. Analogous to the $\lambda_t$'s, we may also construct the right-regular representation $t \to \rho_t$, defined by $(\rho_t\xi)(s) = \xi(st)$ (or, equivalently, $\rho_t\xi_s = \xi_{st^{-1}}$ for all $s,t$ in $G$). It is trivial to verify that each $\rho_t$ commutes with each $\lambda_s$, and hence $\{\rho_t\colon t \in G\} \subseteq M'$. Since $\rho_t\Omega = \xi_{t^{-1}}$ for every $t$, it is clear that $\Omega$ is cyclic for $M'$, as asserted. (It is, in fact, the case that $M' = \{\rho_t\colon t \in G\}''$; this will emerge in the next section, as a consequence of the Tomita-Takesaki theorem, towards which landmark we shall now head.)

## 2.3. The Tomita-Takesaki Theorem (For States)

Recall that a faithful normal tracial positive linear functional on a factor will, by restriction, yield a dimension function which assigns a finite value to 1. Hence, an infinite factor does not admit such a functional. However, every factor (operating on a separable Hilbert space) does admit several faithful normal states [cf. Example 2.2.3(c)].

Suppose, then, that $\phi$ is a faithful normal state on a (not necessarily factorial) von Neumann algebra $M$. According to Ex. (2.2.7), $\phi$ is tracial if and only if $\|\pi_\phi(x)\Omega_\phi\| = \|\pi_\phi(x^*)\Omega_\phi\|$ for all $x$ in $M$. So, in order to study infinite factors, it might be instructive to examine (the lack of isometry of) the operator $\pi_\phi(x)\Omega_\phi \to \pi_\phi(x^*)\Omega_\phi$. The advisability of such an investigation is convincingly demonstrated by the celebrated Tomita-Takesaki theorem, which provides a powerful tool for the study of infinite factors (and factors of type III, in particular).

Till further notice, assume that $M$ is a von Neumann algebra acting on $\mathcal{H}$ and that $\Omega$ is a cyclic and separating vector for $M$.

**Proposition 2.3.1.** *Let $S_0$ and $F_0$ be the conjugate-linear operators, with domains $M\Omega$ and $M'\Omega$, respectively, defined (unambiguously) by $S_0(x\Omega) = x^*\Omega$, $F_0(x'\Omega) = x'^*\Omega$. Then $S_0$ and $F_0$ are densely defined closable operators; their closures, denoted by $S$ and $F$, respectively, satisfy $S = F^* = F_0^*$ and $F = S^* = S_0^*$.*

**Proof.** Since $\Omega$ is cyclic and separating for $M$ as well as for $M'$ (cf. Ex. (2.2.6)), it is clear that both $S_0$ and $F_0$ are densely and unambiguously defined. If $x \in M$ and $x' \in M'$, observe that

$$\begin{aligned} \langle F_0(x'\Omega), x\Omega\rangle &= \langle x'^*\Omega, x\Omega\rangle \\ &= \langle \Omega, x'x\Omega\rangle \\ &= \langle \Omega, xx'\Omega\rangle \\ &= \langle x^*\Omega, x'\Omega\rangle \\ &= \langle S_0(x\Omega), x'\Omega\rangle. \end{aligned}$$

By definition of the adjoint of a conjugate-linear operator, this says that $M\Omega \subseteq \text{dom } F_0^*$ and that $F_0^*|M\Omega = S_0$; i.e., $S_0 \subseteq F_0^*$. This implies that $F_0^*$ is densely defined and hence $F_0$ is closable. If $F$ denotes the closure of $F_0$, then $F^* = F_0^* \supseteq S_0$. So $S_0$ is closable and if $S$ denotes the closure of $S_0$, then $F^* = F_0^* \supseteq S$.

For the reverse inclusion, suppose $\xi \in \text{dom } F_0^*$ and $F_0^*\xi = \xi^\#$. Define operators $Q_0$ and $Q_0^+$, both with domain $M'\Omega$ by $Q_0(x'\Omega) = x'\xi$ and $Q_0^+(x'\Omega) = x'\xi^\#$. Notice that if $x', y' \in M'$, then

$$\begin{aligned} \langle Q_0(x'\Omega), y'\Omega\rangle &= \langle x'\xi, y'\Omega\rangle \\ &= \langle \xi, x'^*y'\Omega\rangle \\ &= \langle \xi, F_0(y'^*x'\Omega)\rangle \\ &= \langle y'^*x'\Omega, \xi^\#\rangle \quad \text{(by definition of } \xi^\#) \\ &= \langle x'\Omega, Q_0^+(y'\Omega)\rangle. \end{aligned}$$

Hence, as before, $Q_0^* \supseteq Q_0^+$ and consequently $Q_0$ is closable; if $Q$ denotes the closure of $Q_0$, then $Q^* = Q_0^* \supseteq Q_0^+$. It is trivial to verify that if $x' \in M'$, then $Q_0x' \supseteq x'Q_0$, and hence, by an easy approximation argument, we may pass to the closure and conclude that the closed densely defined linear operator $Q$ is affiliated to $M$.

Let $Q = u|Q|$ be the polar decomposition of $Q$. Then, $u \in M$ and $e_n \in M$ where $e_n = 1_{[0,n]}(|Q|)$ for each $n$. Also

$$q_n = Qe_n = u(|Q|1_{[0,n]}(|Q|)) \in M$$

(cf. Ex. (1.1.4)). Note that $e_n \nearrow$ rp $Q^*$ and $ue_nu^* \nearrow$ rp $Q$; since $\xi = Q\Omega$ and $\xi^\# = Q^*\Omega = Q^+\Omega$, conclude that $ue_nu^*\xi \to \xi$ and $e_n\xi^\# \to \xi^\#$. On the other hand,

$$q_n\Omega = ue_n|Q|\Omega = ue_nu^*u|Q|\Omega = ue_nu^*Q\Omega = ue_nu^*\xi,$$

while $S_0(q_n\Omega) = q_n^*\Omega = e_nQ^*\Omega = e_n\xi^\#$. Conclude, from the definition of $S$, that $\xi \in$ dom $S$ and $S\xi = \xi^\#$, thereby establishing that $F_0^* \subseteq S$, and hence, that $S = F_0^* = F^*$.

The dual statements, with $F$ and $S$ interchanged, are proved identically, with the roles of $M$ and $M'$ interchanged. □

**Proposition 2.3.2.** *Let $S = J\Delta^{1/2}$ be the polar decomposition of the closed operator $S$. Then,*

(a) *$J$ is a self-adjoint antiunitary operator and $\Delta$ is a positive self-adjoint operator which is invertible in the sense of unbounded operators* (i.e., $\Delta$ *is* 1-1);
(b) *$F = J\Delta^{-1/2}$ is the polar decomposition of $F$; further $\Delta = FS$ and $\Delta^{-1} = SF$;*
(c) $J\Delta J = \Delta^{-1}$; more generally, if $f$ is any (possibly unbounded but finite-valued) Borel function on $[0,\infty)$, then $Jf(\Delta)J = \bar{f}(\Delta^{-1})$; in particular $J\Delta^{it}J = \Delta^{it}$ for all $t \in \mathbb{R}$.

**Proof.** By definition, we have $\Delta = (\Delta^{1/2})^2 = S^*S = FS$ and so $\Delta$ is positive; also $J$ is a conjugate linear partial isometry with initial space $\overline{\text{ran } F}$ and final space $\overline{\text{ran } S}$. Since $M'\Omega = \text{ran } F_0 \subseteq \text{ran } F$ and $M\Omega = \text{ran } S_0 \subseteq \text{ran } S$, conclude that $J$ is antiunitary (i.e., its initial and final spaces are both $\mathcal{H}$).

Since $S_0 = S_0^{-1}$, a simple approximation argument shows that $S$ is invertible and $S = S^{-1}$; this implies that $\Delta$ is invertible since ker $\Delta^{1/2}$ = ker $S = \{0\}$; the equation $S = S^{-1}$ also implies that $J\Delta^{1/2} = S = S^{-1} = \Delta^{-1/2}J^*$. Hence $J^2\Delta^{1/2} = J\Delta^{-1/2}J^*$. Since $J\Delta^{-1/2}J^*$ is an invertible positive self-adjoint operator, the uniqueness of the polar decomposition guarantees that $J^2 = 1$ and $\Delta^{1/2} = J\Delta^{-1/2}J^*$. As $J$ is antiunitary, this yields $J = J^{-1} = J^*$ and $J\Delta^{-1/2}J = \Delta^{1/2}$. This completes the proof of (a) as well as of the decisive step in the proof of (c). The validity of (c) can be derived from $J\Delta^{-1/2}J = \Delta^{1/2}$ and a routine application of the spectral theory to $\Delta^{1/2}$; the verification is purely routine in nature and may be safely left to the reader.

As for (b), conjugate the equation $S = J\Delta^{1/2}$ to get $F = \Delta^{1/2}J = JJ\Delta^{1/2}J = J\Delta^{-1/2}$; this completes the proof. □

The stage is now set for the Tomita-Takesaki theorem. The proof of this powerful and difficult theorem is quite long and elaborate. As stated in the introduction to this chapter, we shall only supply the proof in the very special case when $S$ (or equivalently $\Delta$) is bounded. For a reasonably short proof of the general assertion, the reader may consult [BRI].

**Theorem 2.3.3.** *With the notation established in the preceding propositions, the following statements are valid:*

(a) $\Delta^{it}M\Delta^{-it} = M$ *for all* $t$ *in* $\mathbb{R}$;
*and*
(b) $JMJ = M'$.

**Proof.** Suppose that $S$ is bounded. This means that $S$, $\Delta$ and $F$ are everywhere defined bounded operators; it also means that $\Delta^{-1}$ is an everywhere defined bounded operator, since $\Delta^{-1} = J\Delta J$.

For $x,y$ and $z$ in $M$, note that

$$(1) \qquad ((SxS)y)(z\Omega) = (Sx)(z^*y^*\Omega) = yzx^*\Omega.$$

Apply this with $y = 1$ to conclude that $(SxS)(z\Omega) = zx^*\Omega$; combined with (1), this yields: $(SxS)y)(z\Omega) = (y(SxS))(z\Omega)$. Infer from the cyclicity of $\Omega$ that $(SxS)y = y(SxS)$. Since $x$ and $y$ were arbitrary, we get: $SMS \subseteq M'$.

An entirely analogous argument, with the roles of $(M,S)$ and $(M',F)$ interchanged, yields: $FM'F \subseteq M$.

A combination of the two inclusions established above results in $\Delta M\Delta^{-1} \subseteq M$ (since $\Delta = FS$ and $\Delta^{-1} = SF$); conclude that

$$(2) \qquad \Delta^n M\Delta^{-n} \subseteq M \quad \text{for } n = 1,2, \ldots .$$

For $z \in \mathbb{C}$, let $g(\lambda) = \lambda^z = e^{z \log \lambda}$ for $\lambda \in (0,\infty)$, and write $\Delta^z = g(\Delta)$. Since $\Delta$ is invertible, it is clear that $\Delta^z$ is bounded, for each $z$ in $\mathbb{C}$. Now, fix $x \in M$, $x' \in M'$, $\xi,\eta \in \mathcal{H}$ and consider the (clearly entire) function defined by

$$f(z) = \|\Delta\|^{-2z} \langle [\Delta^z x\Delta^{-z}, x']\xi,\eta\rangle, \; z \in \mathbb{C}$$

where $[a,b]$ denotes the commutator defined by $[a,b] = ab - ba$. A crude estimation yields

$$|f(z)| \leqslant \|\Delta\|^{-2 \operatorname{Re} z} 2\|\Delta^z\| \cdot \|\Delta^{-z}\| \cdot \|x\| \cdot \|x'\| \cdot \|\xi\| \cdot \|\eta\|.$$

However, according to Proposition 2.3.2 (c), we have $J\Delta^z J = \Delta^{-\bar{z}} = (\Delta^{-z})^*$, and consequently $\|\Delta^z\| = \|\Delta^{-z}\|$; thus

$$\|\Delta^z\| \cdot \|\Delta^{-z}\| = \|\Delta^z\|^2 = \|(\Delta^z)^*\Delta^z\| = \|\Delta^{2 \operatorname{Re} z}\| = \|\Delta\|^{2 \operatorname{Re} z},$$

if Re $z \geqslant 0$.

Thus, the entire function $f$ is bounded (by $2\|x\| \|x'\| \|\xi\| \|\eta\|$) in the half-plane Re $z \geqslant 0$; also, by (2), $f(n) = 0$ for $n = 0,1,2, \ldots$. It follows from Carlson's theorem (cf. [Tit]) -- which is valid under a weaker condition on the growth of $f$ ($f(z) = 0(e^{k|z|})$ for Re $z \geqslant 0$, with some $k < \pi$, to be brutally precise) -- that $f(z) = 0$ for all $z$ in $\mathbb{C}$.

Conclude that $\Delta^z M \Delta^{-z} \subseteq M$ for all $z$ in $\mathbb{C}$; applying this to $(-z)$ yields $\Delta^z M \Delta^{-z} = M$ for all $z$ in $\mathbb{C}$; for $z = it$, this is precisely the contention of (a).

It also follows that $\Delta^z M' \Delta^{-z} = M'$ for all $z$ in $\mathbb{C}$; hence,

$$JMJ = J\Delta^{1/2} M \Delta^{-1/2} J = SMS \subseteq M'$$

and

$$JM'J = J\Delta^{-1/2} M' \Delta^{-1/2} J = FM'F \subseteq M,$$

thus establishing (b). □

Before proceeding with some of the consequences of the Tomita-Takesaki Theorem, it would help to get some elementary statements out of the way. So, here come some more exercises.

## Exercises

**(2.3.4)** Let Aut $M$ denote the group of *-automorphisms of $M$.

(a) If $\alpha \in$ Aut $M$, then $\alpha$ is an isometric $\sigma$-weak homeomorphism of $M$ onto itself. (Hint: $\alpha$ preserves the order structure in $M_+$, as does $\alpha^{-1}$; conclude that $\alpha$ is normal. Appeal to the proof of the second half of Theorem 2.2.1.)

(b) Suppose $M \subseteq \mathfrak{L}(\mathcal{H})$ and suppose $\{u_t\}_{t \in \mathbb{R}}$ is a strongly continuous one-parameter group of unitary operators on $\mathcal{H}$ such that $u_t M u_t^* = M$ for all $t$. Show that the equation $\alpha_t(x) = u_t x u_t^*$, $t \in \mathbb{R}$, $x \in M$, defines a one-parameter group of *-automorphisms of $M$ which is $\sigma$-strongly* continuous in the sense that $t_n \to t \Rightarrow \alpha_{t_n}(x) \to \alpha_t(x)$ ($\sigma$-strongly*) for each $x$ in $M$. (Hint: if $x \in M$, the set $\{\alpha_t(x): t \in \mathbb{R}\}$ is norm-bounded, so (cf. Ex. (0.3.4) (d)) it suffices to prove (strong* and hence) strong continuity; this requires a standard $\epsilon/3$ argument.) □

Returning to the setting of the Tomita-Takesaki theorem, it follows from the above exercise that $x \to \Delta^{it} x \Delta^{-it}$ defines a $\sigma$-strongly* continuous one-parameter group of *-automorphism of $M$.

Suppose now that $M$ is a general von Neumann algebra and that $\phi$ is a faithful normal positive linear functional on $M$ with associated GNS triple $(\mathcal{H}_\phi, \pi_\phi, \Omega_\phi)$. Then $\Omega_\phi$ is a cyclic and separating vector for

$\pi_\phi(M)$. Here and in the sequel, we shall denote by $S_\phi$, $F_\phi$, $J_\phi$ and $\Delta_\phi$ the operators arising, in this case, from the considerations that led to the Tomita-Takesaki theorem.

**Definition 2.3.5.** With the above notation, the operators $J_\phi$ and $\Delta_\phi$ are called, respectively, the modular conjugation and the modular operator associated with the pair $(M,\phi)$. Denote by $\{\sigma_t^\phi : t \in \mathbb{R}\}$ the $\sigma$-weakly continuous one-parameter group of $^*$-automorphisms of $M$ defined by

$$\sigma_t^\phi(x) = \pi_\phi^{-1}(\Delta_\phi^{it}\, \pi_\phi(x)\Delta_\phi^{-it}).$$

(The continuity assertion follows from Ex. 2.3.4 (b) and the fact that $\pi_\phi$ is a $\sigma$-weak homeomorphism.) The one-parameter group $\{\sigma_t^\phi\}$ is called the group of modular automorphisms (or simply, the modular group) associated with the positive functional $\phi$. □

**Exercises**

**(2.3.6)** Let $\phi$ be a faithful normal positive linear functional on $M$, and $\mathcal{H}_\phi$, $\pi_\phi$, $\Delta_\phi$, $S_\phi F_\phi$, $J_\phi$ and $\Delta_\phi$ be as above.

(a) Show that $\Omega_\phi \in \text{dom } S_\phi \cap \text{dom } F_\phi$ and $S_\phi\Omega_\phi = F_\phi\Omega_\phi = \Omega_\phi$; conclude that $\Omega_\phi \in \text{dom } \Delta_\phi$ and $\Delta_\phi\Omega_\phi = J_\phi\Omega_\phi = \Omega_\phi$;

(b) Show that $\pi_\phi(\sigma_t^\phi(x))\Omega_\phi = \Delta_\phi^{it}\pi_\phi(x)\Omega_\phi$.

(c) Show that the following conditions are equivalent:

(i) $\phi$ is tracial;
(ii) $S_\phi$ is antiunitary;
(iii) $S_\phi = J_\phi = F_\phi$;
(iv) $\Delta_\phi = 1_{\mathcal{H}_\phi}$;
(v) $\sigma_t^\phi(x) = x \quad \forall x \in M, t \in \mathbb{R}$.

(Hint: The implications (i) $\Leftrightarrow$ (ii) $\Leftrightarrow$ (iii) $\Leftrightarrow$ (iv) $\Rightarrow$ (v) are easy; use (b) for (v) $\Rightarrow$ (iv).) □

Let us now look at some examples that will illustrate the Tomita-Takesaki theorem.

**Example 2.3.7.** (a) Let $M = L^\infty(X,\mathcal{F},\mu)$ and $\phi(f) = \int f\, d\nu$, where $\nu$ is a finite measure equivalent to $\mu$. The GNS triple is given by $\mathcal{H}_\phi = L^2(X,\mu)$, $\pi_\phi(f) = m_f$ and $\Omega_\phi = (d\nu/d\mu)^{1/2}$. It is easy to see that $S_\phi = J_\phi = F_\phi$ is the map $f \to \bar{f}$ on $L^2(X,\mu)$ and that $\Delta_\phi = 1$ -- note that $\phi$ is tracial as $M$ is abelian. Thus, the theorem informs us that $M' = JMJ = M$ (since $Jm_fJ = m_{\bar{f}}$), which is the content of Ex. (0.4.5) (b).

(b) Let $M = \mathcal{L}(\mathcal{H})$ and $\phi(x) = \operatorname{tr} \rho x$, where $\rho$ is an injective positive trace-class operator, with canonical decomposition

$$\rho = \Sigma\, \alpha_n\, t_{\xi_n,\xi_n}\,,$$

where $\alpha_n > 0$, $\Sigma\alpha_n < \infty$ and $\{\xi_n\}$ is an orthonormal basis for $\mathcal{H}$. One version of the GNS triple is given thus:

$$\mathcal{H}_\phi = \ell^2(\mathbb{N}\,;\,\mathcal{H}) = \{\tilde{\xi}: \mathbb{N} \to \mathcal{H}: \sum_n \|\tilde{\xi}(n)\|^2 < \infty\},$$

where $\mathbb{N} = \{1,2,...\}$ (we have assumed $\mathcal{H}$ is infinite dimensional; if not, $\mathbb{N}$ must be replaced by a set of cardinality $\dim \mathcal{H}$); $(\pi_\phi(x)\xi)(n) = x\xi(n)$; $\Omega_\phi(n) = \alpha_n^{1/2}\xi_n$. A natural orthonormal basis for $\mathcal{H}_\phi$ is given by $\{\xi_n^{(m)}: m,n \in \mathbb{N}\}$, where $\tilde{\xi}_n^{(m)}(k) = \delta_{mk}\xi_n$. For $m,n$ in $\mathbb{N}$, define

$$x_n^{(m)} = \alpha_m^{-1/2} t_{\xi_n,\xi_m}$$

and note that $\pi_\phi(x_n^{(m)})\Omega_\phi = \tilde{\xi}_n^{(m)}$, while

$$\pi_\phi(x_n^{(m)*})\Omega_\phi = \left[\frac{\alpha_n}{\alpha_m}\right]^{1/2} \tilde{\xi}_m^{(n)},$$

and thus

$$S_0\tilde{\xi}_n^{(m)} = \left[\frac{\alpha_n}{\alpha_m}\right]^{1/2} \tilde{\xi}_m^{(n)}.$$

It is not too hard to verify (do it!) that the linear span $\mathcal{D}_0$ of $\{\tilde{\xi}_n^{(m)}: m,n \in \mathbb{N}\}$ is contained in dom $F$ and that

$$F\tilde{\xi}_n^{(m)} = \left[\frac{\alpha_m}{\alpha_n}\right]^{1/2} \tilde{\xi}_m^{(n)}.$$

Conclude that $\mathcal{D}_0 \subseteq \operatorname{dom} \Delta$ and that $(\Delta\tilde{\xi})(k) = \alpha_k{}_{-1}\rho\tilde{\xi}(k)$ for all $\tilde{\xi}$ in $\mathcal{D}_0$. Note that $\mathcal{D}_0$ is a core for the positive self-adjoint operator defined on

$$\left\{\tilde{\xi} \in \ell^2(\mathbb{N}\,;\,\mathcal{H}): \sum_k \frac{1}{\alpha_k^2} \|\rho\tilde{\xi}(k)\|^2 < \infty\right\}$$

by $\oplus_k(\alpha_k^{-1}\rho)$, and conclude that, by self-adjointness, $\Delta$ is this operator: $(\Delta\xi)(k) = \alpha_k^{-1}\rho\xi(k)$. (The details are spelt out in Ex. (2.5.5) & (2.5.6).).) Since $S = J\Delta^{1/2}$, conclude that $J\tilde{\xi}_m^{(n)} = \tilde{\xi}_n^{(m)}$ for all $m,n$. It is easily deduced from the above formula for $\Delta$ that $\sigma_t^\phi(x) = \rho^{it}x\rho^{-it}$ for all $x \in M$, $t \in \mathbb{R}$.

In an isomorphic picture, $\mathcal{H}_\phi = \mathcal{H} \otimes \mathcal{H}$, $\pi_\phi(x) = x \otimes 1$, $\Omega_\phi = \Sigma_n\alpha_n^{1/2}\xi_n \otimes \xi_n$, $\Delta_\phi = \rho \otimes \rho^{-1}$ and $J_\phi$ is the unique antiunitary operator such that $J_\phi(\xi_n \otimes \xi_m) = \xi_m \otimes \xi_n$ (this is essentially the flip map, but for the fact that $J_\phi$ is conjugate linear). An easy computation now reveals that $J_\phi(x \otimes 1)J_\phi = 1 \otimes \bar{x}$, where $\bar{x}$ is that operator on $\mathcal{H}$ whose matrix with respect to the orthonormal basis $\{\xi_n\}$ is the (entrywise) complex-conjugate of that of $x$. Hence $J_\phi(\mathcal{L}(\mathcal{H}) \otimes 1)J_\phi = 1 \otimes \mathcal{L}(\mathcal{H})$ and, consequently, $(\mathcal{L}(\mathcal{H}) \otimes 1)' = 1 \otimes \mathcal{L}(\mathcal{H})$.

More generally, it may be shown (see [Tak 1] for details) using the Tomita-Takesaki theorem that if, for von Neumann algebras $M \subseteq \mathcal{L}(\mathcal{H})$ and $N \subseteq \mathcal{L}(\mathcal{K})$, we define $M \otimes N = \{x \otimes y: x \in M, y \in N\}'' \subseteq \mathcal{L}(\mathcal{H} \otimes \mathcal{K})$ then $(M \otimes N)' = M' \otimes N'$. This fact is quite non-trivial and remained an open problem for a long time before Tomita resolved it. We shall, however, not go into a proof of this here.

(c) Let $G$ be a countable group and $M = W^*(G)$, the group von Neumann algebra discussed at the end of Section 2.2. It was shown there that $\Omega = \xi_\epsilon$ is a cyclic and separating vector for $M$. Since $\lambda_t^* \xi_\epsilon = \lambda_{t^{-1}} \xi_\epsilon = \xi_{t^{-1}}$, it follows that the linear span $\mathcal{D}_0$ of $\{\xi_t\}_{t \in G}$ is contained in dom $S$ and that

$$(S\xi)(t) = \overline{\xi(t^{-1})}$$

for all $t$ in $G$ and $\xi$ in $\mathcal{D}_0$ -- recall that $S$ is conjugate linear. Since this map is norm preserving and $\mathcal{D}_0$ is dense in $\ell^2(G)$, conclude that $(S\xi)(t) = \overline{\xi(t^{-1})}$ for all $\xi$ in $\ell^2(G)$, that $J = F = S$ and that $\Delta = 1$ (cf. Ex. (2.3.6)(c)). It follows from the definition of $J$ that $J\lambda_t J = \rho_t$; hence $M' = JMJ = \{\rho_t: t \in G\}''$; thus, the commutant of the left-regular representation is the von Neumann algebra generated by the right-regular representation. □

## 2.4. Weights and Generalized Hilbert Algebras

A positive measure $\mu$ on a Borel space $(X,\mathcal{F})$ can be viewed as a (necessarily faithful and normal) positive linear functional on $L^\infty(X,\mathcal{F},\mu)$, via integration, only when it is a finite measure. If $\mu$ is infinite, some bounded functions -- typically the non-zero constants -- are not $\mu$-integrable. Worse still, no reasonable sense can be made of $\int f\, d\mu$ if, for example, $f = 1_E - 1_F$, where $E$ and $F$ are disjoint sets of infinite measure. With non-negative functions, however, there is no such difficulty and the integral can always be meaningfully defined as a "number" in $[0,\infty]$. These heuristics prompt the following definition.

**Definition 2.4.1.** A weight on a von Neumann $M$ algebra is a mapping $\phi: M_+ \to [0,\infty]$ such that $\phi(\lambda x + y) = \lambda\phi(x) + \phi(y)$ whenever $x,y \in M_+$ and $\lambda \in [0,\infty)$ -- with the convention that $\lambda + \infty = \infty$ and $\lambda \cdot \infty = \infty$ if $\lambda > 0$, while $0 \cdot \infty = 0$. The weight $\phi$ is said to be

(i) faithful, if $x \in M_+$ and $x \neq 0$ imply $\phi(x) > 0$;
(ii) normal, if $\phi(x) = \sup \phi(x_i)$ whenever $x$ is the ($\sigma$-strong*) limit of a monotone increasing net $\{x_i\}$ in $M_+$;
(iii) a trace, if $\phi(x^*x) = \phi(xx^*)$ for all $x$ in $M$. □

**Exercises**

(2.4.2) A weight $\phi$ is said to be finite if $\phi(x) < \infty$ for all $x$ in $M_+$. Prove the equivalence of the following conditions: (i) $\phi$ is finite; (ii) $\phi(1) < \infty$; (iii) there exists $\tilde{\phi} \in M^*_+$ such that $\phi = \tilde{\phi}|M_+$. □

Fundamental to the study of a weight $\phi$ is an analysis of certain subspaces of $M$, whose definitions are given below.

**Definition 2.4.3.** For a weight $\phi$ on $M$, define

$$\mathcal{D}_\phi = \{x \in M_+ : \phi(x) < \infty\} ;$$

$$\mathcal{N}_\phi = \{x \in M : \phi(x^*x) < \infty\} ;$$

$$\mathcal{M}_\phi = \mathcal{N}_\phi^* \mathcal{N}_\phi = \left\{ \sum_{i=1}^{n} x_i^* y_i : x_i, y_i \in \mathcal{N}_\phi, \ n = 1,2, \ldots \right\}.$$

When there is only one weight under discussion, and no confusion seems likely, the subscript $\phi$ will be omitted. □

Clearly, one example of a faithful normal trace (which is not finite) is given by $M = L^\infty(X,\mathcal{F},\mu)$ and $\phi(f) = \int f \, d\mu$, where $\mu$ is an infinite measure. In this case, $\mathcal{D}_\phi = \{f \in L^\infty(X,\mu) : f \geq 0$ a.e. and $f \in L^1(X,\mathcal{F},\mu)\}$, $\mathcal{N}_\phi = L^\infty(X,\mu) \cap L^2(X,\mu)$ and $\mathcal{M}_\phi = L^\infty(X,\mu) \cap L^1(X,\mu)$. Note, in this case, that $\mathcal{D}_\phi = \mathcal{M}_\phi \cap M_+$; this equality is always valid and will soon be established. Before that, however, it will be good to see an example where $M$ is not abelian. Prior to discussing the canonical trace on $\mathcal{L}(\mathcal{H})$, it might be prudent to collect some basic facts about the so-called Hilbert-Schmidt operators; this is done in the following set of exercises.

**Exercises**

(2.4.4)

(a) If $x \in \mathcal{L}(\mathcal{H})$ and $\{\xi_n\}$, $\{\eta_n\}$ are a pair of orthonormal bases for $\mathcal{H}$, prove that $\Sigma\|x\xi_n\|^2 = \Sigma\|x\eta_n\|^2$ -- i.e., if one side is finite, so is the other, and the two sides are equal. (Hint:

$$\Sigma\|x\xi_n\|^2 = \Sigma\Sigma|<x\xi_n,\eta_m>|^2 = \Sigma\|x^*\eta_m\|^2;$$

apply this with $\eta_n = \xi_n$ to get $\Sigma\|x\xi_n\|^2 = \Sigma\|x^*\xi_n\|^2$; combine this with the previous equality applied to $x^*$.)

(b) If $x \in \mathcal{L}(\mathcal{H})$, define $\|x\|_2^2 = \Sigma\|x\xi_n\|^2$, where $\{\xi_n\}$ is any orthonormal basis for $\mathcal{H}$. Let $C_2(\mathcal{H}) = \{x \in \mathcal{L}(\mathcal{H}) : \|x\|_2 < \infty\}$; the members of $C_2(\mathcal{H})$ are called Hilbert-Schmidt operators. Prove the following:

(i) $\|x\| \leq \|x\|_2$ for $x \in C_2(\mathcal{H})$. (Hint: if $\xi$ is a unit vector, consider an orthonormal basis $\{\xi_n\}$ such that $\xi_1 = \xi$.)
(ii) $\|\cdot\|_2$ is a norm on $C_2(\mathcal{H})$ with respect to which $C_2(\mathcal{H})$ is a Hilbert space. (Hint: use (i) to locate the limit of a Cauchy sequence in $C_2(\mathcal{H})$; the inducing inner-product is given by $\langle x,y\rangle = \Sigma \langle x\xi_n, y\xi_n\rangle$, where $\{\xi_n\}$ is any orthonormal basis for $\mathcal{H}$.)
(iii) $x \in C_2(\mathcal{H}) \Rightarrow x^* \in C_2(\mathcal{H})$ and $\|x\|_2 = \|x^*\|_2$. (Hint: see hint to (a).)
(iv) $C_2(\mathcal{H})$ is a two-sided ideal in $\mathfrak{L}(\mathcal{H})$; in fact, if $x \in C_2(\mathcal{H})$ and $y \in \mathfrak{L}(\mathcal{H})$, then $\|yx\|_2 \leq \|y\| \, \|x\|_2$ and $\|xy\|_2 \leq \|y\| \, \|x\|_2$. (Hint: the first inequality is easy, and, together with (iii), it implies the second.)
(v) Let $x \geq 0$. Then $x \in C_2(\mathcal{H}) \Leftrightarrow x \in K(\mathcal{H})$ and $x$ admits a decomposition $x = \Sigma \, \alpha_n t_{\xi_n,\xi_n}$, where $\alpha_n \geq 0$, $\Sigma \, \alpha_n^2 < \infty$ and $\{\xi_n\}$ is an orthonormal sequence. (Hint: for $\Leftarrow$, extend $\{\xi_n\}$ to an orthonormal basis and use that basis to compute $\|x\|_2$; for $\Rightarrow$, use $\|x\|_2 < \infty$ to conclude that, for each $\epsilon > 0$, $1_{(\epsilon,\infty)}(x)$ has finite rank, and hence that $x$ is compact.)
(vi) If $x \in \mathfrak{L}(\mathcal{H})$, show that $x \in C_2(\mathcal{H})$ if and only if $x^*x \in \mathfrak{L}(\mathcal{H})_*$, in which case $\|x\|_2^2 = \mathrm{tr}\ x^*x$. (Hint: if $x \in C_2(\mathcal{H})$ has polar decomposition $x = u|x|$, use (v) to conclude $|x|^2 \in \mathfrak{L}(\mathcal{H})_*$; if

$$|x|^2 = \Sigma \, \alpha_n \, t_{\xi_n,\xi_n}$$

with $\alpha_n \geq 0$, $\Sigma \, \alpha_n < \infty$ and $\{\xi_n\}$ orthonormal, extend $\{\xi_n\}$ to an orthonormal basis for $\mathcal{H}$ and compute $\|x\|_2$.)
(vii) Let $x \in \mathfrak{L}(\mathcal{H})_+$. The following conditions are equivalent: ($\alpha$) $x \in \mathfrak{L}(\mathcal{H})_*$; ($\beta$) $\Sigma \langle x\xi_n,\xi_n\rangle < \infty$ for every orthonormal basis $\{\xi_n\}$; ($\gamma$) $\Sigma\langle x\xi_n,\xi_n\rangle < \infty$ for some orthonormal basis $\{\xi_n\}$. (Hint: for ($\gamma$) $\Rightarrow$ ($\alpha$), consider $x^{1/2}$). □

Let $M = \mathfrak{L}(\mathcal{H})$, with $\mathcal{H}$ (separable and) infinite-dimensional. Define $\phi\colon \mathfrak{L}(\mathcal{H})_+ \to [0,\infty]$ by $\phi(x) = \Sigma\langle x\xi_n,\xi_n\rangle$ where $\{\xi_n\}$ is any orthonormal basis for $\{\xi_n\}$. Thus $\phi(x) = \mathrm{tr}\ x$ if $x$ is of trace class and $\phi(x) = \infty$, otherwise (by Ex. (2.4.4) (b) (vii)) -- in particular, the definition is independent of $\{\xi_n\}$. It is clear that $\phi$ is a faithful trace; the fact that $\phi$ is a trace follows from Ex. (2.4.4) (b) (iii). (In fact, the reason for using the word trace, in the sense of Definition 2.4.1 (iii), stems from this example.) Furthermore, $\phi$ is normal. The verification is fairly easy; if $x_i \nearrow x$, the cases $x \in \mathfrak{L}(\mathcal{H})_*$ and $x \notin \mathfrak{L}(\mathcal{H})_*$ must be treated separately. Note, finally, that for this $\phi$, $N_\phi = C_2(\mathcal{H})$ and $M_\phi = \mathfrak{L}(\mathcal{H})_*$.

Let us proceed now with the analysis of a general weight.

**Proposition 2.4.5.** *Let $\phi$ be a weight on $M$, with associated spaces $\mathfrak{D}$, $N$ and $M$ as in Definition* 2.4.3.

(a) $\mathcal{D}$ *is a hereditary positive cone; i.e.,* $x,y \in \mathcal{D}$ *and* $\lambda \in [0,\infty) \Rightarrow \lambda x + y \in \mathcal{D}$, *and* $x \in \mathcal{D}$, $z \in M_+$, $z \leqslant x \Rightarrow z \in \mathcal{D}$;
(b) $\mathcal{N}$ *is a left-ideal in* $M$;
(c) $\mathcal{M}$ *is a self-adjoint subalgebra of* $M$ *(not necessarily containing* 1 *and not necessarily closed in any topology)*;
(d) $\mathcal{D} = \mathcal{M}_+ = \mathcal{M} \cap M_+$, *and every element of* $\mathcal{M}$ *is a linear combination of four elements of* $\mathcal{D}$;
(e) $x,z \in \mathcal{N}$, $y \in M \Rightarrow x^*yz \in \mathcal{M}$;
(f) *there is a unique linear functional* $\dot{\phi}$ *on* $\mathcal{M}$ *such that* $\dot{\phi}|\mathcal{M}_+ = \phi$.

**Proof.** (a) Easy.

(b) Trivially $\mathcal{N}$ is closed under scalar multiplication. If $x,y \in M$, note that

$$(x+y)^*(x+y) \leqslant (x+y)^*(x+y) + (x-y)^*(x-y)$$

$$= 2(x^*x + y^*y);$$

conclude that $\mathcal{N} + \mathcal{N} \subseteq \mathcal{N}$. Finally, if $x,y \in M$, note that $(yx)^*(yx) = x^*y^*yx \leqslant \|y\|^2 x^*x$, and infer that $M\mathcal{N} \subseteq \mathcal{N}$.

(c) This follows easily from (b), since $\mathcal{N}^*$ is a right-ideal.

(d) Suppose $z = \Sigma_{j=1}^{n} x_j^* y_j$, with $x_j, y_j \in \mathcal{N}$. Apply the polarization identity to get

$$4z = \sum_{j=1}^{n} \sum_{k=0}^{3} i^k (y_j + i^k x_j)^*(y_j + i^k x_j)$$

which proves the second assertion in (d), since, by (b),

$$\sum_{j=1}^{n} (y_j + i^k x_j)^*(y_j + i^k x_j) \in \mathcal{D}, \qquad \text{for each } k = 0,1,2,3.$$

If $z \in \mathcal{M}_+$, then $z = z^*$ and hence,

$$4z = 2(z + z^*)$$

$$= \sum_{j=1}^{n} 2(x_j^* y_j + y_j^* x_j)$$

$$= \sum_{j=1}^{n} \{(x_j + y_j)^*(x_j + y_j) - (x_j - y_j)^*(x_j - y_j)\}$$

$$\leqslant \sum_{j=1}^{n} (x_j + y_j)^*(x_j + y_j) \in \mathcal{D};$$

conclude from (a) that $z \in \mathcal{D}$; this establishes that $\mathcal{M}_+ \subseteq \mathcal{D}$. Conversely, if $x \in \mathcal{D}$, then $x \geqslant 0$ and $x^{1/2} \in \mathcal{N} \cap \mathcal{N}^*$, so that $x \in \mathcal{M}_+$.

(e) This follows at once from (b).

(f) Uniqueness of such a $\dot{\phi}$ is an immediate consequence of (d). Existence is also a consequence of (d): use additivity of $\phi$ on $\mathcal{D}$ to extend in a well-defined fashion to the real linear span of $\mathcal{M}_+$, and then extend to all of $\mathcal{M}$. (This is exactly the way that, after having defined the Lebesgue integral for non-negative functions, one extends the notion of the integral to complex integrable functions.) □

A weight $\phi$ can be trivial in the following sense:

$$\phi(x) = \begin{cases} 0, & \text{if } x = 0, \\ \infty, & \text{if } x \underset{\neq}{\geq} 0 \end{cases}$$

For this $\phi$, $\mathfrak{D}_\phi = \mathfrak{N}_\phi = \mathfrak{M}_\phi = \{0\}$ and not much more can be said.

**Definition 2.4.6.** A weight $\phi$ on $M$ is said to be semifinite if $\mathfrak{M}_\phi$ is $\sigma$-weakly dense in $M$. □

Loosely speaking, semifiniteness means that there are sufficiently many elements at which $\phi$ has a finite value. For example, if $M = L^\infty(X,\mathfrak{F},\mu)$ and $\phi(f) = \int f\, d\nu$, where $\nu$ is a positive measure with the same null sets as $\mu$, then (under the standing assumption of $\sigma$-finiteness of $\mu$), semifiniteness of $\nu$ is equivalent to $\sigma$-finiteness of $\nu$. Observe, also, that the canonical trace on $\mathfrak{L}(\mathcal{H})$ is semifinite, since $\mathfrak{L}(\mathcal{H})_*$, by virtue of containing all finite rank operators, is $\sigma$-weakly dense in $\mathfrak{L}(\mathcal{H})$. In the following exercises, some alternative characterizations of semifiniteness are given, which say that semifiniteness is the same as ampleness of $\mathfrak{D}$ in one sense or another.

**Exercises**

(2.4.7)

(a) If $h,k \in M_+$ satisfy $h \leq k$ and if $h$ is invertible (i.e., $h^{-1} \in \mathfrak{L}(\mathcal{H})$), then $k$ is invertible and $k^{-1} \leq h^{-1}$. (Hint: Observe that if $x \geq 0$, then (by an easy application of the spectral theorem) $x$ is invertible iff there exists $\epsilon > 0$ such that $x \geq \epsilon.1$; this takes care of the first assertion. For the second, $h \leq k \Rightarrow h^{-1/2}hh^{-1/2} \leq h^{-1/2}kh^{-1/2}$, and so $\|k^{1/2}h^{-1/2}\xi\| \geq \|\xi\|$ for all $\xi$; if $x \in \mathfrak{L}(\mathcal{H})$ is invertible, with polar decomposition $x = u|x|$, then $x^* = |x|u^* = u^*xu^*$, conclude that $\|h^{-1/2}k^{1/2}\xi\| \geq \|\xi\|$ for all $\xi$, whence $1 \leq k^{1/2}h^{-1}k^{1/2}$, and so $k^{-1} = k^{-1/2}1k^{-1/2} \leq h^{-1}$.)

(b) If $h \in M_+$ and $\epsilon > 0$, define $h_\epsilon = h(1 + \epsilon h)^{-1}$. Verify that

(i) $h_\epsilon \geq 0$ and $h_\epsilon \in M$;
(ii) $h_\epsilon \leq h_{\epsilon'}$ if $\epsilon \geq \epsilon'$ and $h_\epsilon \nearrow h$ as $\epsilon \downarrow 0$;
(iii) $h \leq k \Rightarrow h_\epsilon \leq k_\epsilon$. (Hint: Use (a) to get $(1 + \epsilon h)^{-1} \geq (1 + \epsilon k)^{-1}$, and note that

$$h_\epsilon = \frac{1}{\epsilon}(1 - (1 + \epsilon h)^{-1}).)$$

**(2.4.8)** Let $\phi$ be a weight on $M$ and let $\mathfrak{D}$, $\mathfrak{N}$ and $\mathfrak{M}$ be as usual. Let $\Lambda = \{x \in \mathfrak{D}: \|x\| < 1\}$.

(a) Show that $\Lambda$ is directed upwards; i.e., if $x_1, x_2 \in \Lambda$, there exists $x \in \Lambda$ such that $x_1 \leqslant x$ and $x_2 \leqslant x$. (Hint: Verify that $h_i = x_i(1-x_i)^{-1} \in \mathfrak{D}$, since $h_i \leqslant kx_i$ for some $k > 0$, and that $x_i = h_i(1+h_i)^{-1}$. Let $h = h_1 + h_2$ ($\in \mathfrak{D}$), put $x = h(1+h)^{-1}$ and use Ex. (2.4.7) (b) (iii).)

(b) Since, by (a), $\Lambda$ may be viewed as a monotone increasing net in $M$, let $\bar{x} = \lim \Lambda$. Show that $\bar{x}$ is the projection $\bar{e} = \vee\{e: e \in \mathcal{P}(M) \cap \mathfrak{D}\}$. (Hint: If $e \in \mathcal{P}(M) \cap \mathfrak{D}$ and $0 < \lambda < 1$, then $\lambda e \in \Lambda$ and so $\lambda e \leqslant \bar{x}$; hence $e \leqslant \bar{x}$. Since $0 \leqslant \bar{x} \leqslant 1$, conclude that $e \leqslant 1_{\{1\}}(\bar{x})$, and so $\bar{e} \leqslant 1_{\{1\}}(\bar{x})$. Conversely, if $x \in \Lambda$, argue that $1_{[\epsilon,\infty)}(x) \in \mathfrak{D}$ and so $1_{(0,\infty)}(x) \leqslant \bar{e}$; thus $x = x\bar{e}$; conclude that $\bar{x} = \bar{x}\,\bar{e} = \bar{e}$.)

(c) With $\bar{e}$ as above, show that the $\sigma$-weak closure of $\mathfrak{M}$ is $\bar{e}\,M\,\bar{e}$. (Hint: If $x \in \Lambda$, then $x \leqslant \bar{x} = \bar{e}$ so $x \in \bar{e}M\bar{e}$. Conclude that the $\sigma$-weak closure of $\mathfrak{M}$ is contained in $\bar{e}M\bar{e}$. For the converse, if $y \in M_+$, then $x \in \Lambda \Rightarrow x^{1/2}yx^{1/2} \in \mathfrak{D}$ by Prop. 2.4.5 (e). Since $\Lambda$ is a bounded net, argue that

$$\lim_{x \in \Lambda} x^{1/2}yx^{1/2} = \bar{e}y\bar{e}\ ,$$

in the $\sigma$-weak sense and so $\bar{e}y\bar{e}$ lies in the $\sigma$-weak closure of $\mathfrak{M}$.

(d) The following conditions on a weight $\phi$ are equivalent:

(i) $\phi$ is semifinite
(ii) $1 = \vee\{e \in \mathcal{P}(M): \phi(e) < \infty\}$
(iii) there is an increasing net $\{x_i\}$ in $\mathfrak{D}$ such that $\|x_i\| < 1$ for all $i$ and $x_i \nearrow 1$. □

Concerning normality of weights, there is a result due to Haagerup which establishes the equivalence of several definitions of normality. We state the result below with no proof, but we shall use it freely in the future. (For the reader who would like to see the sort of "topological-vector-space" arguments that one can never really get away from in this subject, it might be instructive to see the original proof, in [Haa].)

**Proposition 2.4.9.** *For a weight $\phi$ on $M$, the following conditions are equivalent:*

(i) *$\phi$ is normal;*
(ii) *there exists a monotone increasing net $\{\psi_i: i \in I\}$ in $M_{*,+}$ such that $\psi_i(x) \nearrow \phi(x)$ for all $x$ in $M_+$;*
(iii) *there exists a family $\{\psi_i: i \in J\} \subseteq M_{*,+}$ such that $\phi(x) = \Sigma_{i\in J}\psi_i(x)$ for all $x$ in $M_+$ (the sum being interpreted as the limit of the net of finite sums);*

(iv) *$\phi$ is $\sigma$-weakly lower semicontinuous; i.e., if $x_i \to x$ $\sigma$-weakly, $x_i, x \in M_+$, then* $\phi(x) \leq \liminf \phi(x_i)$. □

With very minor modifications, the GNS construction goes through for weights.

**Proposition 2.4.10.** *Let $\phi$ be a faithful, normal, semifinite weight on $M$. Let $\mathfrak{D}_\phi$, $\mathfrak{N}_\phi$ and $\mathfrak{M}_\phi$ be the associated subspaces of $M$, as in Definition 2.4.3. Let us use the same symbol $\phi$ for the extension, as a linear functional, to $\mathfrak{M}_\phi$, as in Prop. 2.4.5 (f). Then there exists a triple $(\mathcal{H}_\phi, \pi_\phi, \eta_\phi)$ where*

(i) *$\mathcal{H}_\phi$ is a Hilbert space;*
(ii) *$\pi_\phi$ is a *-algebra homomorphism of $M$ into $\mathfrak{L}(\mathcal{H}_\phi)$;*
(iii) *$\eta_\phi$: $\mathfrak{N}_\phi \to \mathcal{H}_\phi$ is a linear map such that*

$$\langle \eta_\phi(x), \eta_\phi(y) \rangle = \phi(y^*x), \quad \pi_\phi(z)\eta_\phi(x) = \eta_\phi(zx)$$

*whenever $x, y \in \mathfrak{N}_\phi$ and $z \in M$, and such that $\eta_\phi(\mathfrak{N}_\phi)$ is dense in $\mathcal{H}_\phi$.*

*The triple is unique in the sense that if $(\mathcal{H}', \pi', \eta')$ is another such triple, there exists a unique unitary operator $u$: $\mathcal{H}_\phi \to \mathcal{H}'$ such that $u\eta_\phi(x) = \eta'(x)$ for all $x$ in $\mathfrak{N}_\phi$ and $\pi'(z) = u\pi_\phi(z)u^*$ for every $z$ in $M$.*

*Furthermore, $\pi_\phi$ is isometric and is a $\sigma$-weak homeomorphism of $M$ onto $\pi_\phi(M)$.*

**Proof.** The proof is a repetition of the proof for finite weights, with only minor and obvious modifications. It will suffice to start the reader off on the proof by suggesting that $\mathcal{H}_\phi$ be taken as the completion of $\mathfrak{N}_\phi$ with respect to the inner product given by $\langle x,y \rangle = \phi(y^*x)$, and that the fact that $\mathfrak{N}_\phi$ is a left ideal must be periodically recalled. (Note that this is implicit in the statement (iii).) □

For the sake of brevity and convenience of exposition, we shall henceforth write 'fns' for the cumbersome expression 'faithful, normal and semifinite'. Suppose $\phi$ is a fns weight on $M$, with associated spaces $\mathfrak{D}$, $\mathfrak{N}$ and $\mathfrak{M}$, and GNS triple $(\mathcal{H}, \pi, \eta)$. Since $\pi$ is an isomorphism, we shall identify $M$ with $\pi(M)$ and assume $M \subseteq \mathfrak{L}(\mathcal{H})$, $\pi(x) = x$. Let $\mathfrak{U} = \eta(\mathfrak{N} \cap \mathfrak{N}^*)$. If $\xi_i = \eta(x_i) \in \mathfrak{U}$, $i = 1,2$, (recalling that the faithfulness of $\phi$ implies the injectivity of $\eta$), write $\xi_1\xi_2 = \eta(x_1x_2)$ and $\xi_1^\# = \eta(x_1^*)$.

**Proposition 2.4.11.**

(a) *is an involutive, associative algebra;*
(b) *is equipped with an inner product which satisfies:*
   (i) *$\langle \xi\eta, \zeta \rangle = \langle \eta, \xi^\#\zeta \rangle$ for all $\xi, \eta, \zeta$ in $\mathfrak{U}$;*
   (ii) *for each $\xi$ in $\mathfrak{U}$, the map $\eta \to \xi\eta$ is a continuous linear operator on $\mathfrak{U}$, with respect to the inner product;*

(c) $\mathcal{U}^2$ $\left(= \left\{ \sum_{i=1}^{n} \xi_i\eta_i : \xi_i, \eta_i \in \mathcal{U},\ n = 1,2, \ldots \right\}\right)$ *is dense in* $\mathcal{U}$;

*and*

(d) *the conjugate linear operator* $S_0$: $\mathcal{U} \to \mathcal{U}$, *defined by* $S_0\xi = \xi^\#$, *extends to a closed operator* $S$ *in the completion* $\mathcal{H}$ *of* $\mathcal{U}$.

**Proof.** The assertions (a) and (b) are clear.

(c) Since $\phi$ is semidefinite, there exists a monotone increasing net $\{x_i\}$ such that $x_i \in \mathcal{D}$, $\|x_i\| < 1$ and $x_i \nearrow 1$ (cf. Ex. (2.4.8) (d)). For each $i$, note that $x_i^{1/2} \in N \cap N^*$; if $x \in N$, also $x_i^{1/2}x \in N^*N \subseteq N \cap N^*$, since $N$ (resp. $N^*$) is a left- (resp., right-) ideal in $M$. Since $x_i \to 1$ strongly (in fact, even $\sigma$-strongly), conclude that if $\xi = \eta(x)$, with $x \in N$, then

$$\xi = \lim_i x_i\xi = \lim_i \xi_i\eta_i ,$$

with $\xi_i = \eta(x_i^{1/2})$ and $\eta_i = \eta(x_i^{1/2}x)$. This shows that $\mathcal{U}^2$ is in fact dense in $\eta(N)$, which is more than what (c) states; note, in particular that $\mathcal{U}$ is dense in $\mathcal{H}$.

The proof of (d) is somewhat involved, and is outlined in the following exercises. □

## Exercises

**(2.4.12)** (This exercise is the GNS construction for elements of $M_{*,+}$, which are not necessarily faithful.) Let $\psi \in M_{*,+}$.

(a) Show that $I_\psi = \{x \in M: \psi(x^*x) = 0\}$ is a left-ideal in $M$. (Hint: imitate the proof of the corresponding statement regarding $N_\phi$.)

(b) The equation $\langle x + I_\psi, y + I_\psi \rangle = \psi(y^*x)$ defines (unambiguously!) an inner-product on the vector space $M/I_\psi$.

(c) Let $\mathcal{H}_\psi$ be the completion of $M/I_\psi$ with respect to the above inner product, and let $\Omega_\psi = 1 + I_\psi$. Show that there exists a unique normal *-homomorphism $\pi_\psi$: $M \to \mathcal{L}(\mathcal{H}_\psi)$ such that $\pi_\psi(x)(y+I_\psi) = xy + I_\psi$ for all $x,y$ in $M$. Note that $\pi_\psi(M)$ is a von Neumann algebra of operators on $\mathcal{H}_\psi$ and that $\psi(x) = \langle \pi_\psi(x)\Omega_\psi, \Omega_\psi \rangle$ for all $x$ in $M$.

(d) Formulate and prove a uniqueness statement concerning the GNS triple $(\mathcal{H}_\psi, \pi_\psi, \Omega_\psi)$.

**(2.4.13)** Let $\phi$ be a fns weight on $M$; assume that $M \subseteq \mathcal{L}(\mathcal{H})$ and that $\pi_\phi(x) = x$ for $x$ in $M$; also, write $\eta$ for $\eta_\phi$. Suppose $\psi \in M_{*,+}$ satisfies $\psi \leq \phi$ (i.e., $\psi(x) \leq \phi(x)$ for all $x$ in $M_+$).

(a) Show that there exists a unique operator $a' \in M'$ such that $0 \leq a' \leq 1$ and $\psi(y^*x) = \langle a'\eta(x), \eta(y) \rangle$ for all $x,y$ in $N$ (= $N_\phi$). (Hint: note that the sesquilinear form $[\eta(x),\eta(y)] = \psi(y^*x)$ is bounded (by 1), and use density of $\eta(N)$ in $\mathcal{H}$ to lay hands on $a'$.)

(b) Let $(\mathcal{H}_\psi, \pi_\psi, \Omega_\psi)$ be a GNS triple for $\psi$. Show that there exists a unique isometric operator $u$: $\mathcal{H}_\psi \to \mathcal{H}$ such that $u\ \pi_\psi(x)\Omega_\psi = a'^{1/2}\eta(x)$ for all $x$ in $\mathcal{N}$, with $a'$ as in (a). (Hint: note that $\mathcal{N} \supseteq \mathcal{M}$, so that (by semifiniteness of $\phi$) $\mathcal{N}$ is $\sigma$-weakly dense in $M$; argue that $\pi_\psi(\mathcal{N})\Omega_\psi$ must be dense in $\mathcal{H}_\psi$; observe that

$$\langle \pi_\psi(x)\Omega_\psi, \pi_\psi(y)\Omega_\psi \rangle = \psi(y^*x) = \langle a'^{1/2}\eta(x), a'^{1/2}\eta(y) \rangle$$

for all $x,y$ in $\mathcal{N}$.)

(c) In the notation of (b), show that $u\ \pi_\psi(x) = xu$ for all $x$ in $M$; conclude that if $\xi_\psi = u\Omega_\psi$, then

(i) $x\xi_\psi = a'^{1/2}\eta(x)$ for $x$ in $\mathcal{N}$;
(ii) if $\{x_i\}$ is any net in $\mathcal{N}$ such that $x_i \to 1$ strongly, then $\xi_\psi = \lim a'^{1/2}\eta(x_i)$, and consequently (i) uniquely determines $\xi_\psi$;
(iii) $\psi(x) = \langle x\xi_\psi, \xi_\psi \rangle$ for all $x$ in $M$. (Hint: verify this for $x$ in $\mathcal{N}^*\mathcal{N}$, using (i) and the definition of $a'$, and note that both sides of the desired equation depend $\sigma$-weakly continuously on $x$.)

(iv) $[M\xi_\psi] = \overline{\operatorname{ran} a'}$. (Hint: $[M\xi_\psi] = [\mathcal{N}\xi_\psi] = [a'^{1/2}\eta(\mathcal{N})]$.)

**(2.4.14)** Let $\phi$ be a fns weight on $M$. Let $\{\psi_i: i \in I\}$ be a monotone increasing net in $M_{*,+}$ such that $\psi_i(x) \nearrow \phi(x)$ for each $x$ in $M_+$ (cf. Prop. 2.4.9 (ii)). For each $i \in I$, let $a_i' \in M_+'$ and $\xi_i$ $(= \xi_{\psi_i})$ be associated with $\psi_i$ as in Ex. (2.4.13) (a) and (c), respectively.

(a) $\{a_i': i \in I\}$ is a monotone increasing net in $M_+'$ and $a_i' \nearrow 1$.
(b) Show that $\{\xi_i: i \in I\}$ is cyclic and separating for $M$, and hence, also for $M'$. (Hint: if $x \in M$, $\langle x\xi_i, \xi_i \rangle = \psi_i(x) \nearrow \phi(x)$, and $\phi$ is faithful; so $\{\xi_i\}$ is separating for $M$. By Ex. (2.4.13) (iv), $[M\xi_i] = \overline{\operatorname{ran} a_i'}$; also $a_i' \nearrow 1$ by (a).)
(c) The set $S = \{a_i'^{1/2} a'\xi_j: i,j \in I, a' \in M'\}$ is total in $\mathcal{H}$, i.e., $[S] = \mathcal{H}$. (Hint: use (a) and (b).)
(d) If $S_0$ is as in Prop. 2.4.11 (d), show that the set $S$ of (c) is contained in dom $S_0^*$; so, $S_0^*$ is densely defined and $S_0$ is closable. (Hint: use Ex. (2.4.13) (c) (i) repeatedly to show that, for any $x \in \mathcal{N} \cap \mathcal{N}^*$, $a' \in M'$ and $i,j \in I$, $\langle S_0\eta(x), a_i'^{1/2}a'\xi_j \rangle = \langle a_j'^{1/2}a'^*\xi_i, \eta(x) \rangle$.) □

**Definition 2.4.15.** An involutive algebra $(\mathcal{U}, \#)$, equipped with an inner product, which satisfies the conditions (a) - (d) of Prop. 2.4.11 is called a generalized Hilbert algebra. It is called a Hilbert algebra if (d) is strengthened (considerably!) to the requirement that $S_0$ is isometric -- i.e., if $\|\xi^\#\| = \|\xi\|$ for all $\xi$ in $\mathcal{U}$. □

Remark that if $\mathcal{U} = \eta_\phi(\mathcal{N}_\phi \cap \mathcal{N}_\phi^*)$ arises from a fns weight $\phi$ on $M$, then $\mathcal{U}$ is a Hilbert algebra if and only if $\phi$ is a trace. If an abstract generalized Hilbert algebra $\mathcal{U}$ is given, we shall let $\mathcal{H}$ denote the completion of $\mathcal{U}$, and we shall write $S$ and $F$, respectively, for the closure of $S_0$ and the adjoint of $S_0$ respectively. These are sometimes referred to as the "sharp" and "flat" operators, respectively. We shall have a word to say at a later stage about the appropriateness of the usage of these musical terms.

We shall write $\mathcal{D}^{\#}$ and $\mathcal{D}^{b}$ for the domains of $S$ and $F$, respectively, and write $\xi^{\#} = S\xi$ and $\eta^{b} = F\eta$ whenever $\xi \in \mathcal{D}^{\#}$ and $\eta \in \mathcal{D}^{b}$. (Notice the obvious fact that if $\xi \in \mathcal{U}$, this is consistent with the "sharp" operation in $\mathcal{U}$.)

It is easy to see that, given a generalized Hilbert algebra $\mathcal{U}$ with completion $\mathcal{H}$, there is a map $\pi$: $\mathcal{U} \to \mathfrak{L}(\mathcal{H})$ such that $\pi(\xi)\eta = \xi\eta$ for all $\xi,\eta$ in $\mathcal{U}$, and that, further, the map $\pi$ is a homomorphism of involutive algebras, i.e., $\pi(\xi\eta) = \pi(\xi)\pi(\eta)$ and $\pi(\xi^{\#}) = \pi(\xi)^*$. The density of $\mathcal{U}^2$ in $\mathcal{U}$ translates into non-degeneracy (in the sense of the word as used in the double-commutant theorem) of the self-adjoint operator algebra $\pi(\mathcal{U})$.

Consider, now, the following class of "right-bounded" elements:

$$\mathcal{U}' = \{\eta \in \mathcal{D}^{b}; \exists\, c > 0 \ni \|\pi(\xi)\eta\| \leqslant c\|\xi\| \quad \forall\xi \text{ in } \mathcal{U}\}.$$

Each $\eta$ in $\mathcal{U}'$ clearly gives rise to a bounded operator $\pi'(\eta)$ on $\mathcal{H}$ such that $\pi'(\eta)\xi = \pi(\xi)\eta$ for all $\xi$ in $\mathcal{U}$. It is easily checked that $\pi'(\mathcal{U}') \subseteq \pi(\mathcal{U})'$ -- the old principle of left multiplications commuting with right multiplications. A simple argument -- of the sort repeatedly used in the proof of Prop. 2.3.1 -- shows that if $\eta \in \mathcal{U}'$, then $\eta^{b} \in \mathcal{U}'$ and $\pi'(\eta^{b}) = \pi'(\eta)^*$. The following assertions are also valid: $\mathcal{U}'$ is dense in $\mathcal{H}$ -- in fact $\mathcal{U}'$ is a core for $F$, meaning $\mathcal{U}' \subseteq \mathcal{D}^{b}$ and the graph of $F|\mathcal{U}'$ is dense in the graph of $F$; $\pi'(\mathcal{U}')$ is a self-adjoint subalgebra of $\mathfrak{L}(\mathcal{H})$ -- in fact, if $\eta_1,\eta_2 \in \mathcal{U}'$, then $\pi'(\eta_2)\eta_1 \in \mathcal{U}'$ and $\pi'(\pi'(\eta_2)\eta_1) = \pi'(\eta_1)\pi'(\eta_2)$; finally, $\pi'(\mathcal{U}')$ is strongly dense in $\pi(\mathcal{U})'$.

Continue this game one step further and consider the set

$$\mathcal{U}'' = \{\xi \in \mathcal{D}^{\#}: \exists\, c > 0 \ni \|\pi'(\eta)\xi\| \leqslant c\|\eta\| \quad \forall\eta \in \mathcal{U}'\}.$$

It is clear that $\mathcal{U} \subseteq \mathcal{U}''$, but the inclusion may be strict -- suppose $\mathcal{U} = C[0,1]$, regarded as a subset of $L^2[0,1]$; then both $\mathcal{U}'$ and $\mathcal{U}''$ are equal to $L^\infty[0,1]$.

**Definition 2.4.16.** The generalized Hilbert algebra $\mathcal{U}$ is said to be achieved if $\mathcal{U} = \mathcal{U}''$; the von Neumann algebra $\pi(\mathcal{U})''$ is called the left von Neumann algebra of $\mathcal{U}$, and denoted by $\mathfrak{L}(\mathcal{U})$. □

It is not too hard to show that the generalized Hilbert algebra arising from a fns weight on a von Neumann algebra $M$ is achieved; if $\mathcal{U} = \eta_\phi(\mathcal{N}_\phi \cap \mathcal{N}_\phi^*)$, then the left von Neumann algebra $\mathfrak{L}(\mathcal{U})$ is just $\pi_\phi(M)$. In the converse direction, Combes has shown (cf. [Com 2]) that every achieved generalized Hilbert algebra $\mathcal{U}$ is of the form

$\mathcal{U} = \eta_\phi(\mathcal{N}_\phi \cap \mathcal{N}_\phi^*)$, where, in fact, $M = \mathfrak{L}(\mathcal{U})$ and $\phi$ is defined by

$$\phi(x) = \begin{cases} \|\xi\|^2, & \text{if } x = \pi(\xi^\# \xi), \text{ with } \xi \in \mathcal{U} \\ \infty, & \text{if } x \text{ is not of the above form.} \end{cases}$$

If $\mathcal{U}$ is not achieved, one can argue with $\mathcal{U}''$ exactly as one argued with $\mathcal{U}'$ to find that if, for $\xi$ in $\mathcal{U}''$, one lets $\pi(\xi)$ denote the unique operators on $\mathcal{H}$ such that $\pi(\xi)\eta = \pi'(\eta)\xi$ for all $\eta$ in $\mathcal{U}'$, then $\pi(\mathcal{U}'')$ is a self-adjoint algebra of operators on $\mathcal{H}$. This, in turn, equips $\mathcal{U}''$ with the structure of a generalized Hilbert algebra, which can be shown to be achieved; finally, one has the fact that $\pi'(\mathcal{U}')' = \pi(\mathcal{U})'' = \pi(\mathcal{U}'')''$, so that both the generalized Hilbert algebras $\mathcal{U}$ and $\mathcal{U}''$ have the same left von Neumann algebra.

We are now in a position to state the Tomita-Takesaki theorem in its full generality; we shall state it as a theorem about generalized Hilbert algebras, and feel free to interpret it as a statement concerning the GNS representation of a von Neumann algebra that is associated with a fns weight.

**Theorem 2.4.17.** *Let $M = \mathfrak{L}(\mathcal{U})$ be the left von Neumann algebra of a generalized Hilbert algebra $\mathcal{U}$. Let $\mathcal{H}$ denote the completion of $\mathcal{U}$. Then there exist a self-adjoint antiunitary operator $J$ and an invertible (possibly unbounded) positive self-adjoint operator $\Delta$ in $\mathcal{H}$ such that:*

(a) *$S = J\Delta^{1/2}$ and $F = J\Delta^{-1/2}$ are the polar decompositions of the sharp and flat operators;*
(b) *$J\Delta J = \Delta^{-1}$, and consequently $Jf(\Delta)J = \bar{f}(\Delta^{-1})$ for any Borel function $f$ on $(0,\infty)$; and*
(c) *$\Delta^{it}M\,\Delta^{-it} = M$ for all $t \in \mathbb{R}$, and $JMJ = M'$.* □

Having balked at proving the theorem even in the case of a finite weight, we shall (naturally?) say nothing about the proof other then that the reader may find it, as also the preceding facts concerning generalized Hilbert algebras in [Tak 1].

It is implicit in the statement of (a) that $\mathfrak{D}^\# = \text{dom } \Delta^{1/2}$ and $\mathfrak{D}^b = \text{dom } \Delta^{-1/2}$; we rest the case for justifying the "musical" notation "sharp" and "flat".

As in the case of finite weights, if $\phi$ is a fns weight on a von Neumann algebra $M$, with GNS triple $(\mathcal{H}_\phi, \pi_\phi, \eta_\phi)$, an application of the Tomita-Takesaki Theorem to $\pi_\phi(M)$ -- which is the left von Neumann algebra of $\mathcal{U}_\phi = \eta_\phi(\mathcal{N}_\phi \cap \mathcal{N}_\phi^*)$ -- results in a $\sigma$-weakly continuous one-parameter group $\{\sigma_t^\phi\}$ of "modular" automorphisms of $M$, defined by

$$\sigma_t^\phi(x) = \pi_\phi^{-1}(\Delta_\phi^{it}\, \pi_\phi(x) \Delta_\phi^{-it}).$$

**Example 2.4.18.** Let $G$ be a locally compact group with a fixed left Haar measure, which we shall simply denote by $ds$. Recall that the modular function of $G$ is a continuous homomorphism, usually denoted by $\Delta$, from $G$ into the multiplicative group $\mathbb{R}_+^*$ of positive reals, which is characterized by the requirement that, for any $\xi$ in $C_c(G)$ (the space of compactly supported continuous functions on $G$), $\int \xi(st)ds = \Delta(t)^{-1}\int \xi(s)ds$, for every $t$ in $G$. The function $\Delta$ is the Radon-Nikodym derivative of the inversion map in $G$; i.e., $\int \xi(s^{-1})ds = \int \xi(s)\Delta(s^{-1})ds$ for $\xi$ in $C_c(G)$. The left- and right-regular representations of $G$ are the strongly continuous unitary representations of $G$ in $L^2(G,ds)$ defined, respectively, by $(\lambda_t\xi)(s) = \xi(t^{-1}s)$ and $(\rho_t\xi)(s) = \Delta(t)^{1/2}\xi(st)$.

The set $\mathfrak{U} = C_c(G)$ has the structure of a generalized Hilbert algebra, if the product, sharp and inner-product are defined, respectively, by

$$(\xi\eta)(s) = \int \xi(t)\eta(t^{-1}s)dt,$$

$$\xi^{\#}(s) = \Delta(s)^{-1}\ \overline{\xi(s^{-1})},$$

and $$<\xi,\eta> = \int \xi(s)\ \overline{\eta(s)}\ ds\ .$$

The completion $\mathcal{H}$ is just $L^2(G)$, and it is clear, from the definition, that if $\xi \in \mathfrak{U}$, $\pi(\xi) = \int \xi(s)\lambda_s ds$ -- this integral being interpreted weakly $(<\pi(\xi)\eta,\zeta> = \int \xi(s)<\lambda_s\eta,\zeta>ds)$ or in the Bochner sense (since $\xi \in L^1(G)$). One can conclude from the above integral representation that $\mathfrak{L}(\mathfrak{U}) = \{\lambda_t\colon t \in G\}''$. (The inclusion $\subseteq$ is clear, while $\supseteq$ requires the use of standard arguments involving approximate identities.) A few computations reveal that the operators $J$ and $\Delta$ are defined by

$$(J\xi)(s) = \Delta(s)^{-1/2}\ \overline{\xi(s^{-1})} \quad \text{and} \quad (\Delta\xi)(s) = \Delta(s)\xi(s)$$

(the $\Delta$ on the right denoting the modular function).

Using these explicit formulae, it is easily verified that $J\lambda_t J = \rho_t$ for each $t$ in $G$; hence, as in the case of discrete $G$, we have $\{\lambda_t\colon t \in G\}' = \{\rho_t\colon t \in G\}''$. The other half of the Tomita-Takesaki theorem -- $\Delta^{it}M\Delta^{-it} = M$ -- focuses on the commutation relations between multiplication (by $\Delta$) and translation operators.

## 2.5. The KMS Boundary Condition

As seen in the last section, every fns weight $\phi$ on $M$ gives rise to a group $\{\sigma_t^{\phi}\colon t \in \mathbb{R}\}$ of automorphisms. While this group was obtained by passing to the GNS space of $(M,\phi)$ and then back again to $M$, it is useful, in practice, to have an intrinsic description of the modular group associated with a weight, which does not appeal to the GNS construction at all. Such a characterization of the modular group is given by the Kubo-Martin-Schwinger (KMS, in the sequel) boundary condition, which we shall first discuss for finite weights.

**Definition 2.5.1.** A $\phi$ in $M_{*,+}$ is said to satisfy the KMS boundary condition (at inverse temperature $\beta = 1$) with respect to a $\sigma$-weakly continuous one-parameter group $\{\alpha_t: t \in \mathbb{R}\}$ of automorphisms of $M$ (i.e., $\alpha_t(x) \to x$ $\sigma$-weakly as $t \to 0$, for each $x$) if, for each $x$ and $y$ in $M$, there exists a bounded continuous function $F: \{\lambda \in \mathbb{C}: 0 \leqslant \text{Im}\,\lambda \leqslant 1\} \to \mathbb{C}$, which is analytic in the interior of the strip and satisfies

$$F(t + i) = \phi(x\alpha_t(y)) \text{ and } F(t) = \phi(\alpha_t(y)x), \quad \text{for all } t \text{ in } \mathbb{R}. \ \square$$

For brevity in exposition, let us say that "$F$ is KMS-admissible for $x$ and $y$", whenever $F$, $x$ and $y$ are as in Definition 2.5.1. (To be precise, $\phi$ and $\{\alpha_t\}$ must also be incorporated into the abbreviated phrase, but these will be clearly determined by the contexts in which the phrase will be used.)

The following Exercises list a few simple facts from complex function theory, which will smooth out some of the subsequent proofs.

**Exercises**

(2.5.2) Let $t_1 < t_2 < t_3$ be real numbers. For $i = 1,2$, let $f_i: \{\lambda \in \mathbb{C}: t_i \leqslant \text{Im}\,\lambda \leqslant t_{i+1}\} \to \mathbb{C}$ be continuous functions, analytic in the interior of the strip. If $f_1$ and $f_2$ agree on the common boundary $\text{Im}\,\lambda = t_2$, show that the global function consequently defined on $\{\lambda \in \mathbb{C}: t_1 \leqslant \text{Im}\,\lambda \leqslant t_3\}$ is analytic. (Hint: use Morera's converse of Cauchy's theorem.)

(2.5.3) Let $f: \{\lambda \in \mathbb{C}: 0 \leqslant \text{Im}\,\lambda \leqslant 1\} \to \mathbb{C}$ be continuous and analytic in the interior.

(a) If $f$ is real on the real line, show that the function $\tilde{f}: \{\lambda \in \mathbb{C}: -1 \leqslant \text{Im}\,\lambda \leqslant 1\} \to \mathbb{C}$ defined by

$$\tilde{f}(\lambda) = \begin{cases} f(\lambda), & \text{if } 0 \leqslant \text{Im}\,\lambda \leqslant 1 \\ \overline{f(\overline{\lambda})} & \text{if } -1 \leqslant \text{Im}\,\lambda \leqslant 0 \end{cases}$$

is analytic in the interior of the strip.

(b) If $f(t) = f(t + i)$ for all $t$ in $\mathbb{R}$, then $f$ extends uniquely to an entire funtion with period $i$.

(c) If $f$ is constant on either bounding edge, then $f$ is identically constant. (Hint: assume the constant is zero, and apply (a).)

**Lemma 2.5.4.** *If $\phi$ (in $M_{*,+}$) satisfies the KMS condition with respect to (the $\sigma$-weakly continuous one-parameter group) $\{\alpha_t\}$, then $\phi \circ \alpha_t = \phi$ for all $t$.*

**Proof.** Let $F$ be KMS-admissible for 1 and $y$. Then $F(t) = F(t + i) =$

$\phi(\alpha_t(y))$ for all $t$. So, (by Ex. (2.5.3)(b)) the function $F$ extends to an entire function of period $i$. Since, however, $f$ is bounded on the strip $0 \leqslant \text{Im } \lambda \leqslant 1$ (by definition), it follows from Liouville's theorem that $F$ is constant. Consequently, for any $t$,

$$\phi(\alpha_t(y)) = F(t) = F(0) = \phi(y). \qquad \square$$

Since we shall, henceforth, have to deal frequently with unbounded operators, it would be useful -- both for immediate and later needs -- to gather together some easy facts concerning them.

## Exercises

**(2.5.5)** Let $A$ be a densely defined closed operator in $\mathcal{H}$ with domain $\mathfrak{D}$. Recall that a linear subspace $\mathfrak{D}_0$ of $\mathfrak{D}$ is called a core for $A$ if the graph of $A|\mathfrak{D}_0$ is dense in the graph of $A$ (with respect to the norm-topology in $\mathcal{H} \oplus \mathcal{H}$).

(a) If $A = u|A|$ is the polar decomposition of $A$, show that a linear subspace $\mathfrak{D}_0$ is a core for $A$ if and only if it is a core for $|A|$.

(b) If $A$ is self-adjoint, show that

$$\bigcup_{n=1}^{\infty} (\text{ran } (1_{[-n,n]} (A)))$$

is a core for $g(A)$, where $g$ is any continuous (locally bounded is enough!) function on sp $A$.

(c) If $A$ and $B$ are self-adjoint operators in $\mathcal{H}$, show that $A = B$ if and only if there is a core $\mathfrak{D}_0$ for $A$ such that $A|\mathfrak{D}_o \subseteq B$. (Hint: $B$ is closed and $A$ is the closure of $A|\mathfrak{D}_0$; by self-adjointness, $A \subseteq B \Rightarrow B \subseteq A \Rightarrow A = B$.)

**(2.5.6)** If $H_n$ is a self-adjoint operator in $\mathcal{H}_n$ for $n = 1,2, \ldots$, define an operator $H$ in $\oplus \mathcal{H}_n$ by the following prescription:

$$\text{dom } H = \{\oplus \xi_n : \xi_n \in \text{dom } H_n \quad \forall n, \quad \text{and} \quad \sum_n (\|\xi_n\|^2 + \|H_n\xi_n\|^2) < \infty\}$$

$$H(\oplus \xi_n) = \oplus(H_n\xi_n), \quad \text{if } \oplus \xi_n \in \text{dom } H.$$

(a) Show that $H$ is self-adjoint and that $1_E(H) = \oplus 1_E(H_n)$ for every Borel set $E$ in $\mathbb{R}$;

(b) If, for each $n$, $\mathfrak{D}_n$ is a core for $H_n$, and if

$$\mathfrak{D}_0 = \{\oplus \xi_n\text{: } \xi_n \in \mathfrak{D}_n \quad \forall n, \ \xi_n \neq 0 \text{ for only finitely many } n\},$$

then $\mathfrak{D}_0$ is a core for $H$.

(c) What can you say if it is only assumed that each $H_n$ is closed?

**Definition 2.5.7.** By a flow on $M$ is meant a one-parameter group $\{\alpha_t\}_{t\in\mathbb{R}}$ of automorphisms of $M$ such that $t \to \alpha_t(x)$ is $\sigma$-weakly continuous, for each $x$ in $M$. □

We shall now lead up to the main result of this section, which states that if $\phi$ is a faithful normal positive linear functional on $M$, then $\{\sigma_t^\phi\}$ is the only flow on $M$ with respect to which $\phi$ satisfies the KMS boundary condition.

Suppose, then, that $\phi$ is a fixed faithful normal positive linear functional on $M$ and that $\{\alpha_t\}$ is a flow on $M$ such that $\phi \circ \alpha_t = \phi$ for all $t$ in $\mathbb{R}$. To prove the desired statement, we may assume without loss of generality -- by identifying $M$ with $\pi_\phi(M)$ -- that $\phi(x) = \langle x\Omega,\Omega\rangle$ where $\Omega$ is a vector in $\mathcal{H}$ which is cyclic and separating for $M$. The assumption $\phi \circ \alpha_t = \phi$ implies, then, that there exists a unitary operator $u_t$ on $\mathcal{H}$ such that $u_t x\Omega = \alpha_t(x)\Omega$ for all $x$ in $M$; it is trivially verified that $\{u_t\}$ is a strongly continuous one-parameter group of unitary operators on $\mathcal{H}$. Hence, by Stone's theorem, there exists a unique self-adjoint operator $H$ in $\mathcal{H}$ such that $u_t = e^{itH}$ for all $t$. Let $\mathcal{B}_H$ denote the linear subspace spanned by vectors of the form $f(H)x\Omega$, with $x \in M$ and $f \in C_c^\infty(\mathbb{R})$. In the special case when $\alpha_t = \sigma_t^\phi$, we have $u_t = \Delta^{it}$ (cf. Ex. (2.3.6) (b)) and so $H = \log \Delta$; in this case, we shall simply write $\mathcal{B}$ for $\mathcal{B}_{\log\Delta}$.

**Lemma 2.5.8.**

(a) $\mathcal{B}_H$ *is a core for* $g(H)$, *for any continuous function* $g$ *on* $\mathbb{R}$;
(b) $\mathcal{B}_H \subseteq M\Omega$;
(c) *in case* $\alpha_t = \sigma_t^\phi$, *the subspace* $\mathcal{B}$ *is invariant under the sharp operator* $S$.

**Proof.** (a) Note that $g(H)1_K(H)$ is an everywhere defined bounded operator, for any compact set $K \subseteq \mathbb{R}$; hence $\mathcal{B}_H \subseteq \text{dom } g(H)$. In view of Ex. (2.5.5) (b), it suffices to prove the following: if $\xi = 1_K(H)\xi$ for some compact set $K \subseteq \mathbb{R}$, there exist $\xi_n \in \mathcal{B}_H$ such that $\xi_n \to \xi$ and $g(H)\xi_n \to g(H)\xi$. To see this, first pick $x_n$ in $M$ such that $x_n\Omega \to \xi$; next, select any $f \in C_c^\infty(\mathbb{R})$ such that $f(t) = 1$ for all $t$ in $K$. (For the existence of such an $f$, see, for example [Yos].) If $\tilde{K}$ is a compact set containing the support of $f$, observe that $\xi_n = f(H)x_n\Omega \in \mathcal{B}_H$ for all $n$, that $\xi_n \to f(H)\xi = \xi$ (since $f(H)$ is bounded), and that $g(H)\xi_n \to g(H)f(H)\xi = g(H)\xi$ since $g(H)f(H)$ is bounded. (We have used the fact that

$$f(H)\xi = f(H)1_{\tilde{K}}(H)\xi = 1_{\tilde{K}}(H)\xi = \xi.)$$

(b) Let $\xi = f(H)x\Omega \in \mathcal{B}_H$, with $f \in C_c^\infty(\mathbb{R})$ and $x \in M$. Notice that the Fourier transform of $f$ is in $L^1(\mathbb{R})$ -- this is true of even the larger class of so-called Schwartz functions (cf. [Yos]); consequently the inversion theorem of Fourier analysis is applicable:

$$f(\lambda) = (2\pi)^{-1/2} \int_{-\infty}^{\infty} \hat{f}(t)e^{it\lambda}dt.$$

Hence $\xi = (2\pi)^{-1/2}\int \hat{f}(t)e^{itH}x\Omega\ dt$, the integral being interpreted strongly. On the other hand, the "integral" $(2\pi)^{-1/2}\int\hat{f}(t)\alpha_t(x)dt$ makes sense $\sigma$-weakly and defines an element $y$ of $M$. (Such $\sigma$-weak integrals will be treated rigorously in Section 3.2; the reader, if he feels uneasy at the preceding discussion, may take it on faith that the argument is not specious, and re-visit this proof after having perused Section 3.2.) It follows from the definition of $H$ that $e^{itH}x\Omega = \alpha_t(x)\Omega$ and so $\xi = y\Omega$, as desired.

(c) When $H = \log \Delta$, since the function $g(t) = e^{t/2}$ is certainly continuous, conclude from (a) that $\mathcal{B} \subseteq \text{dom}\ \Delta^{1/2} = \mathcal{D}^{\#}$; if $f \in C_c^{\infty}(\mathbb{R})$ and $x \in M$,

$$\begin{aligned} S(f(\log \Delta)x\Omega) &= J\Delta^{1/2}f(\log \Delta)x\Omega \\ &= \bar{f}(-\log \Delta)J\Delta^{1/2}x\Omega \\ &= \bar{f}(-\log \Delta)x^*\Omega \in \mathcal{B}, \end{aligned}$$

since $t \to \bar{f}(-t)$ is also a $C^{\infty}$-function of compact support. So $S\mathcal{B} \subseteq \mathcal{B}$. In fact, since $S = S^{-1}$, we also have $S\mathcal{B} = \mathcal{B}$. □

## Exercises

Retain the above notation.

**(2.5.9)**

(a) $\xi \in \text{dom}\ e^{t_0H}$ and $t_0 > 0 \Rightarrow \xi \in \text{dom}\ e^{tH}$ for $0 \leq t \leq t_0$; further, for any such $t$,

$$\|e^{tH}\xi\|^2 \leq \|1_{(-\infty,0]}(H)\xi\|^2 + \|1_{(0,\infty)}(H)e^{t_0H}\xi\|^2.$$

(Hint: let $\mu_\xi$ be the measure given by $\mu_\xi(E) = \langle 1_E(H)\xi,\xi\rangle$, and observe that

$$\begin{aligned} \int_{-\infty}^{\infty} (e^{t\lambda})^2 d\mu_\xi(\lambda) &= \int_{(\lambda\leq 0)} e^{2t\lambda}d\mu_\xi(\lambda) + \int_{(\lambda>0)} e^{2t\lambda}d\mu_\xi(\lambda) \\ &\leq \mu_\xi((-\infty,0]) + \int_{(\lambda>0)} e^{2t_0\lambda}d\mu_\xi(\lambda).) \end{aligned}$$

(b) If $\{\xi_n\} \subseteq \text{dom}\ e^{t_0H}$, $t_0 > 0$, and if $\xi_n \to 0$ and $e^{t_0H}\xi_n \to 0$, then $e^{tH}\xi_n \to 0$ for $0 \leq t \leq t_0$. (Hint: use the inequality in (a).)

(c) If $\xi \in \text{dom}\ e^{t_0H}$, $t_0 > 0$, there exist $\xi_n$ in $\mathcal{B}_H$ such that $e^{tH}\xi_n \to e^{tH}\xi$ for $0 \leq t \leq t_0$. (Hint: use Lemma 2.5.8 (a) and (b) above.)

(d) If $\xi \in \mathcal{D}^{\#}$, there exist $\xi_n \in \mathcal{B}$ such that $\Delta^t\xi_n \to \Delta^t\xi$ and $\Delta^t\xi_n^{\#} \to \Delta^t\xi^{\#}$ for $0 \leq t \leq 1/2$. (Hint: use (c) with $H = \log \Delta$ and $t_0 = 1/2$, to choose $\xi_n \in \mathcal{B}$ such that $\Delta^t\xi_n \to \Delta^t\xi$ for $t \in [0,1/2]$. Note that $\xi_n^{\#} = J\Delta^{1/2}\xi_n \to J\Delta^{1/2}\xi = \xi^{\#}$, by the above convergence for $t = 1/2$; also $\Delta^{1/2}\xi_n^{\#} = \Delta^{1/2}J\Delta^{1/2}\xi_n = J\xi_n \to J\xi = \Delta^{1/2}\xi^{\#}$; again appeal to (c) with $\xi_n^{\#}$ in place of $\xi_n$.) □

**Lemma 2.5.10.** *If $\xi \in \mathcal{B}$ and $\eta \in \mathcal{D}^{\#}$, then*

$$\langle \Delta^z\xi, \eta^{\#}\rangle = \langle \eta, \Delta^{1-\bar{z}}\xi^{\#}\rangle \quad \text{for all } z \text{ in } \mathbb{C};$$

*the common value defines an entire function of the complex variable $z$.*

**Proof.** Note that as $f(\lambda) = \lambda^z = e^{z \log \lambda}$ is a continuous function on $(0,\infty)$, it follows from Lemma 2.5.8 (a) and (c) that both $\xi$ and $\xi^{\#}$ belong to dom $\Delta^z$ for all $z$ in $\mathbb{C}$, so that the above expressions are meaningful. Further, by Lemma 2.5.8 (b), $\mathcal{B} \subseteq M\Omega$.

First consider the special case when $\eta$ is also in $\mathcal{B}$. Then, $\xi = x\Omega$ and $\eta = y\Omega$ for some $x$ and $y$ in $M$. Then,

$$\begin{aligned}\langle \Delta^z \xi, \eta^{\#}\rangle &= \langle \Delta^z x\Omega, J\Delta^{1/2}y\Omega\rangle \\ &= \langle \Delta^{1/2}y\Omega, J\Delta^z x\Omega\rangle \quad (\text{since } J = J^*) \\ &= \langle \Delta^{1/2}y\Omega, \Delta^{-\bar{z}} Jx\Omega\rangle \quad (\text{since } Jf(\Delta) = \bar{f}(\Delta^{-1})J) \\ &= \langle y\Omega, \Delta^{1-\bar{z}} x^*\Omega\rangle \quad (\text{since } \Delta^{-1/2}J = S) \\ &= \langle \eta, \Delta^{1-\bar{z}} \xi^{\#}\rangle ,\end{aligned}$$

so that the assertion is valid when $\eta \in \mathcal{B}$.

If $\eta \in \mathcal{D}^{\#}$, use Ex. (2.5.9) (d) to pick $\eta_n \in \mathcal{B}$ such that $\eta_n \to \eta$ and $\eta_n^{\#} \to \eta^{\#}$, use the validity of the desired equality for each $n$, pass to the limit and conclude that $\langle \Delta^z\xi, \eta^{\#}\rangle = \langle \eta, \Delta^{1-\bar{z}}\xi^{\#}\rangle$. The verification of the analyticity of the function is routine, and may be safely left to the reader. □

**Theorem 2.5.11.** *Let $\phi$ be a faithful normal positive linear functional on $M$. The following conditions on a flow $\{\alpha_t\}$ on $M$ are equivalent:*

(i) $\alpha_t = \sigma_t^{\phi}$ *for all $t$;*
(ii) $\phi$ *satisfies the* KMS *condition with respect to $\{\alpha_t\}$.*

**Proof.** (i) $\Rightarrow$ (ii): Let $x,y \in M$ and let $\xi = x\Omega$ and $\eta = y\Omega$. Since $\xi \in \mathcal{D}^{\#}$, use Ex. (2.5.9) (d) to pick $\xi_n$ in $\mathcal{B}$ such that $\Delta^t\xi_n \to \Delta^t\xi$ and $\Delta^t\xi_n^{\#} \to \Delta^t\xi^{\#}$ for $0 \leq t \leq 1/2$. Clearly, then, $\Delta^z\xi_n \to \Delta^z\xi$ and $\Delta^z\xi_n^{\#} \to \Delta^z\xi^{\#}$ for $0 \leq \operatorname{Re} z \leq 1/2$ (since $\Delta^{it}$ is unitary). It is not hard to see, using the inequality in Ex. (2.5.9) (a), that the above sequences converge uniformly in the strip $0 \leqslant \operatorname{Re} z \leqslant 1/2$. Since $f_n(z) = \langle \Delta^{-iz}\xi_n, \eta^{\#}\rangle$ is entire, conclude that the function $F_1(z) = \langle \Delta^{-iz}\xi, \eta^{\#}\rangle$ is continuous in the strip $0 \leqslant \operatorname{Im} z \leqslant 1/2$ and analytic in the interior. The function $F_1$ is also bounded in the strip:

$$|F_1(z)| \leqslant (\sup_{t \in [0,1/2]} \|\Delta^t \xi\|)\|\eta^{\#}\| \leqslant (\|\xi\|^2 + \|\Delta^{1/2}\xi\|^2)^{1/2}\|\eta^{\#}\|$$

(again using the estimate in Ex. (2.5.9) (a)).

By Lemma 2.5.10, it is true that $f_n(z) = \langle \eta, \Delta^{1-i\bar{z}}\xi_n^{\#}\rangle$; notice that $\operatorname{Re}(1 - i\bar{z}) = 1 - \operatorname{Im} z$. Arguing exactly as above, note that $\{f_n\}$ converges uniformly in the strip $1/2 \leqslant \operatorname{Im} z \leqslant 1$ to the function

$F_2(z) = \langle \eta, \Delta^{1-i\bar{z}}\xi^{\#}\rangle$, which is bounded and continuous in the closed strip and analytic in the interior.

Observe that $F_1$ and $F_2$ agree on the line $\operatorname{Im} z = 1/2$, since they are, both, the limit of the sequence $\{f_n\}$. Hence (cf. Ex. (2.5.2)),

$$F(z) = \begin{cases} \langle \Delta^{-iz}\xi, \eta^{\#}\rangle, & 0 \leqslant \operatorname{Im} z \leqslant 1/2 \\ \langle \eta, \Delta^{1-i\bar{z}}\xi^{\#}\rangle, & 1/2 \leqslant \operatorname{Im} z \leqslant 1 \end{cases}$$

defines a bounded continuous function on the strip $0 \leqslant \operatorname{Im} z \leqslant 1$, which is analytic in the interior, and satisfies, for $t \in \mathbb{R}$,

$$\begin{aligned} F(t) &= \langle \Delta^{-it}x\Omega, y^*\Omega\rangle \\ &= \langle \Delta^{-it}x\Omega, y^*\Delta^{-it}\Omega\rangle \quad (\text{since } \Delta^{-it}\Omega = \Omega \ \forall t) \\ &= \langle \Delta^{it}y\Delta^{-it}x\Omega, \Omega\rangle \\ &= \phi(\sigma_t^{\phi}(y)x) \end{aligned}$$

and

$$\begin{aligned} F(t+i) &= \langle y\Omega, \Delta^{1-i(t-i)}x^*\Omega\rangle \\ &= \langle y\Omega, \Delta^{-it}x^*\Omega\rangle \\ &= \langle y\Delta^{-it}\Omega, \Delta^{-it}x^*\Omega\rangle \\ &= \langle x\Delta^{it}y\Delta^{-it}\Omega, \Omega\rangle \\ &= \phi(x\sigma_t^{\phi}(y)); \end{aligned}$$

in other words $F$ is KMS-admissible for $x$ and $y$.

(ii) ⇒ (i): If $\{\alpha_t\}$ is a flow with respect to which $\phi$ satisfies the KMS condition, then $\phi \circ \alpha_t = \phi$ by Lemma 2.5.4. It follows from the discussion preceding Lemma 2.5.8 that there exists a self-adjoint operator $H$ in $\mathcal{H}$ such that $\alpha_t(x)\Omega = e^{itH}x\Omega$ for $x$ in $M$ and $t$ in $\mathbb{R}$. Let $\mathcal{B}_H$ be as in Lemma 2.5.8. We need to show that $\Delta = e^H$; it suffices to show that $\Delta \supseteq e^H|\mathcal{B}_H$, since $\mathcal{B}_H$ is a core for $e^H$ [see Ex. (2.5.5)(c)].

Let $\xi = x\Omega \in \mathcal{B}_H$, $y \in M$. Note that the function given by $G(z) = \langle e^{-izH}x\Omega, y^*\Omega\rangle$ is entire (use Lemma 2.5.8 (a)) and that for $t \in \mathbb{R}$, $G(t) = \langle e^{-itH}x\Omega, y^*\Omega\rangle = \langle x\Omega, e^{itH}y^*\Omega\rangle = \langle x\Omega, \alpha_t(y^*)\Omega\rangle = \phi(\alpha_t(y)x)$. It follows (on applying Ex. (2.5.3) (c) to $F - G$ where $F$ is KMS-admissible for $x$ and $y$ (relative to $\alpha_t$)) that $G(t + i) = \phi(x\alpha_t(y))$ for all $t$ in $\mathbb{R}$. Hence,

$$\begin{aligned}\langle\alpha_t(y)\Omega, x^*\Omega\rangle &= G(t + i)\\ &= \langle e^{-i(t+i)H}x\Omega, y^*\Omega\rangle\\ &= \langle e^{H}x\Omega, e^{itH}y^*\Omega\rangle\\ &= \langle e^{H}x\Omega, \alpha_t(y^*)\Omega\rangle .\end{aligned}$$

Setting $\eta = \alpha_t(y)\Omega$, we find that

$$\langle\eta, \xi^{\#}\rangle = \langle e^{H}\xi, S_0\eta\rangle$$

for all $\eta$ in $M\Omega$. This means that $e^H\xi \in \text{dom } F$ and that $F e^H\xi = S\xi$. Since $F = F^{-1}$, dom $F$ = ran $F$ and so $S\xi \in \text{dom } F$ and $\Delta\xi = FS\xi = e^H\xi$; i.e., $\xi \in \text{dom } \Delta$ and $\Delta\xi = e^H\xi$, and the proof is complete. □

An immediate consequence of Theorem 2.5.11 and Lemma 2.5.4 -- which can also be directly verified using the fact that $\Delta^{it}\Omega = \Omega$ -- is stated below as a corollary for convenience of reference.

**Corollary 2.5.12.** $\phi \circ \sigma_t^\phi = \phi$. □

**Definition 2.5.13.** The fixed point algebra of a flow $\alpha = \{\alpha_t\}_{t\in\mathbb{R}}$ on $M$ is the von Neumann subalgebra, denoted by $M^\alpha$, of $M$ given by

$$M^\alpha = \{x \in M : \alpha_t(x) = x \text{ for all } t\}.$$

If $\phi$ is a faithful normal positive linear functional, we shall write $M^\phi$ for $M^{\sigma^\phi}$; thus,

$$M^\phi = \{x \in M : \sigma_t^\phi(x) = x \quad \text{for all } t\}. \quad □$$

The next result is a very elegant characterization of $M^\phi$ which, among other things, drives home the fact that the modular group $\sigma^\phi$ effectively measures the lack of traciality of $\phi$.

**Corollary 2.5.14.** *Let $\phi$ be a faithful normal positive linear functional on $M$. Let $x \in M$. A necessary and sufficient condition for $x$ to belong to the fixed point algebra $M^\phi$ is that $\phi(xy) = \phi(yx)$ for all $y$ in $M$. In particular, $Z(M) \supseteq M^\phi$.*

**Proof.** If $x \in M^\phi$ and $y \in M$, let $F$ be KMS-admissible for $y$ and $x$.

Then $F(t) = \phi(xy)$ and $F(t+i) = \phi(yx)$, for all $t$ in $\mathbb{R}$. Conclude from Ex. (2.5.3) (c) that $F$ is constant, so that, in particular, $\phi(xy) = \phi(yx)$.

Suppose conversely that $\phi(xy) = \phi(yx)$ for all $y$ in $M$. Fix $y$, and let $F$ be KMS-admissible for $x$ and $y$; note that the hypothesis on $x$ ensures that $F(t) = \phi(\sigma_t^\phi(y)x) = \phi(x\sigma_t^\phi(y)) = F(t+i)$. Argue as in the proof of Lemma 2.5.4 that $F$ extends to an entire function which is bounded and hence constant. So, for any $t$, $\phi(yx) = F(0) = F(-t) = \phi(\sigma_{-t}^\phi(y)x) = \phi(y\sigma_t^\phi(x))$, since $\phi = \phi \circ \sigma_t^\phi$. Thus, $\phi(y(\sigma_t^\phi(x) - x)) = 0$ for all $y$ in $M$. Put $y = (\sigma_t^\phi(x) - x)^*$ and appeal to the assumed faithfulness of $\phi$ to conclude that $\sigma_t^\phi(x) = x$. □

All of the foregoing analysis can be extended to the case of fns weights. The KMS condition is amended as follows.

**Definition 2.5.1'.** A fns weight $\phi$ on $M$ is said to satisfy the KMS boundary condition (at $\beta = 1$) with respect to a flow $\{\alpha_t\}$ on $M$ if

(i) $\phi \circ \alpha_t = \phi$ on $M_+$, for all $t$ in $\mathbb{R}$; and
(ii) for every pair of elements $x$ and $y$ in $N_\phi \cap N_\phi^*$, there exists a bounded continuous function $F$: $\{z \in \mathbb{C}: 0 \leq \operatorname{Im} z \leq 1\} \to \mathbb{C}$, which is analytic in the interior of the strip and satisfies, for all $t$ in $\mathbb{R}$, $F(t) = \phi(\alpha_t(y)x)$, $F(t+i) = \phi(x\alpha_t(y))$. □

Note that if $\phi \circ \alpha_t = \phi$, then $\alpha_t(N_\phi) = N_\phi$ and consequently $\alpha_t(M_\phi) = M_\phi$. So, if $x,y \in N_\phi \cap N_\phi^*$, then $\alpha_t(y) \in N_\phi \cap N_\phi^*$ and so $\alpha_t(y)x$, $x\alpha_t(y) \in M_\phi$ and the expressions $\phi(\alpha_t(y)x)$ and $\phi(x\alpha_t(y))$ are meaningful. Notice, incidentally, that the same symbol $\phi$ is also being used for the linear functional induced by the weight $\phi$ on $M_\phi$. The KMS characterization of the modular group $\sigma^\phi$ now becomes:

**Theorem 2.5.11'.** *If $\phi$ is a* fns *weight on $M$, the following conditions on a flow $\{\alpha_t\}$ on $M$ are equivalent*:

(i) $\alpha_t = \sigma_t^\phi$ *for all $t$ in* $\mathbb{R}$;
(ii) $\phi$ *satisfies the* KMS *condition with respect to* $\{\alpha_t\}$. □

We shall say nothing about a proof of this theorem, which can be found in [Com 2]. Instead, we shall briefly dwell upon one concept that goes into this and other proofs of statements concerning semifinite weights.

The first step in proving the above theorem is to grab hold of a nice subspace $U_0$ of $M$ such that $\eta_\phi(U_0)$ can play the role of the set $B$ used in proving the finite case. Such a $U_0$ must have two virtues: (a) it must be sufficiently ample in $M$, so that approximation arguments are available to us; and (b) $\eta_\phi(U_0)$ must be a core for $\Delta^z$, for every $z$ in $\mathbb{C}$. Such a $U_0$ -- called the Tomita algebra by Takesaki -- is constructed as follows:

Call an element $x$ of $M$ "analytic for $\sigma^\phi$" if there exists a function $F$: $\mathbb{C} \to M$ such that (a) $F$ is $\sigma$-weakly analytic in the sense that

$\psi(F(\cdot))$ is an entire function for every $\psi$ in $M_*$, and (b) $F(t) = \sigma_t^\phi(x)$ for $t$ in $\mathbb{R}$. Such a function, if it exists, is unique since an entire function is determined by its values on the real line; we shall write $\sigma_z^\phi(x)$ for $F(z)$, $z \in \mathbb{C}$.

Let $M_0$ denote the set of $\sigma^\phi$-analytic elements in $M$. It is easily verified that $M_0$ is a self-adjoint subalgebra of $M$ containing 1 and that, for $x,y \in M_0$ and $\lambda$, $z \in \mathbb{C}$, one has:

$$\sigma_z^\phi(\lambda x + y) = \lambda\sigma_z^\phi(x) + \sigma_z^\phi(y),$$

$$\sigma_z^\phi(xy) = \sigma_z^\phi(x)\sigma_z^\phi(y),$$

and

$$\sigma_z^\phi(x^*) = \sigma_{\bar{z}}^\phi(x)^*,$$

If $x \in M$ and $\gamma > 0$, the integral

$$(2\pi\gamma^2)^{-1/2} \int_{\mathbb{R}} \exp\left[\frac{-t^2}{2\gamma^2}\right]\sigma_t^\phi(x)dt$$

converges strongly and $\sigma$-weakly to an element $x_\gamma$ of $M$. (This process of "Gaussian smoothing" is an old friend of probabilists.) It is not hard to establish the following facts:

(i) $x_\gamma \in M_0$ and $\sigma_z^\phi(x_\gamma) = (2\pi\gamma^2)^{-1/2} \int_{\mathbb{R}} \exp\left[-\frac{(t-z)^2}{2\gamma^2}\right]\sigma_t^\phi(x)dt$;

(ii) $x_\gamma \to x$ $\sigma$-weakly as $\gamma \to 0$;

(iii) $x \in \mathfrak{M}_\phi \Rightarrow x_\gamma \in \mathfrak{M}_\phi$ and $\phi(x_\gamma) = \phi(x)$. (Prove this first for $x \in \mathfrak{D}_\phi$ and extend by linearity.)

Consequently $M_0$ is $\sigma$-weakly dense in $M$ and $M_0 \cap \mathfrak{M}_\phi$ is $\sigma$-weakly dense in $\mathfrak{M}_\phi$. It can, further, be shown that both $N_\phi \cap N_\phi^*$ and $\mathfrak{M}_\phi$ are invariant under multiplication (from left or right) by elements of $M_0$. This last statement is proved using a fact about self-adjoint operators, which is stated below as an exercise. This fact might also convince the reader of the plausibility of the fact that if $\mathcal{U}_0 = M_0 \cap N_\phi \cap N_\phi^*$, then $\eta_\phi(\mathcal{U}_0)$ is a core for $\Delta_\phi^z$, for each $z$ in $\mathbb{C}$.

## Exercises

**(2.5.15)** Let $H$ be a positive invertible self-adjoint operator in $\mathcal{H}$. Let $\xi \in \mathcal{H}$ and $t_0 > 0$. The following conditions are equivalent:

(i) $\xi \in \operatorname{dom} H^{-iz}$, for $0 \leqslant \operatorname{Im} z \leqslant t_0$;

(ii) $\xi \in \operatorname{dom} H^{t_0}$;

(iii) there exists a (norm-) bounded (strongly) continuous function $F$: $\{z \in \mathbb{C}: 0 \leqslant \operatorname{Im} z \leqslant t_0\} \to \mathcal{H}$ which is (strongly, or equivalently, weakly) analytic in the interior of the strip and satisfies $F(t) = H^{-it}\xi$ for $t$ in $\mathbb{R}$.

If these equivalent conditions are satisfied, then $F(z) = H^{-iz}\xi$ for $0 \leqslant \text{Im } z \leqslant t_0$. (Hint: (i) $\Rightarrow$ (ii) clearly, while (ii) $\Rightarrow$ (i) by Ex. (2.5.9) (a); for (i) $\Rightarrow$ (iii), set $F(z) = H^{-iz}\xi$; for (iii) $\Rightarrow$ (i), let $\zeta \in \text{dom } H^{t_0}$. Note that the functions $z \to \langle F(z),\zeta\rangle$ and $z \to \langle \xi, H^{i\bar{z}}\zeta\rangle$ are analytic in the strip, continuous on the boundary, and agree on the line $\text{Im } z = 0$. So

$$\langle F(it_0),\zeta\rangle = \langle \xi, H^{t_0}\zeta\rangle .$$

Since $\zeta$ was arbitrary and $H^{t_0}$ is self-adjoint, conclude that $\xi \in \text{dom } H^{t_0}$ and $F(it_0) = H^{t_0}\xi$, and more generally, that $\xi \in \text{dom } H^{-iz}$ and $F(z) = H^{-iz}\xi$.) □

The preceding exercise, most of the material before that (on $M_0$, $\mathcal{U}_0$, etc.) and the following extension of Corollary 2.5.14 may be found in [PT].

**Theorem 2.5.14'.** *Let* $\phi$ *be a* fns *weight on* $M$ *and let* $x \in M$. *The following conditions on* $x$ *are equivalent:*

(i) $x \in M^{\phi}$;
(ii) $x\mathfrak{M}_{\phi} \subseteq \mathfrak{M}_{\phi}$, $\mathfrak{M}_{\phi}x \subseteq \mathfrak{M}_{\phi}$ *and* $\phi(xy) = \phi(yx)$ *for all* $y$ *in* $\mathfrak{M}_{\phi}$. *In particular,* $Z(M) \subseteq M^{\phi}$. □

We shall say nothing about the proof, except that the last statement of the theorem does indeed follow from the implication (ii) $\Rightarrow$ (i) and Prop. 2.4.5 (e), and that the interested reader can find the complete proof in [PT].

## 2.6. The Radon-Nikodym Theorem and Conditional Expectations

Let $(X,\mathcal{F},\mu)$ be a (separable and $\sigma$-finite) measure space and let $M = L^{\infty}(X,\mathcal{F},\mu)$. Every (positive $\sigma$-finite) measure $\nu$ on $(X,\mathcal{F})$, which is absolutely continuous with respect to $\mu$, defines a normal semifinite weight $\psi_{\nu}$ on $M$ via the equation $\psi_{\nu}(f) = \int f \, d\nu$ ($f \in M_+$). Conversely, if $\psi$ is a normal semifinite weight on $M$, then the equation $\nu(E) = \psi(1_E)$ defines a (countably additive) $\sigma$-finite measure $\nu$ on $(X,\mathcal{F})$ -- the $\sigma$-finiteness of $\nu$ being a consequence of the semifiniteness of $\psi$, via Ex. (2.4.8) (d) (ii); it is clear that normality of $\psi$ forces $\psi = \psi_{\nu}$. Hence $\sigma$-finite measures, which are absolutely continuous with respect to $\mu$, are in bijection with semifinite normal weights on $M$.

Notice that $\psi_{\nu}$ is a fns weight on $M$ precisely when $\nu$ is equivalent to $\mu$. Thus, the classical Radon-Nikodym theorem asserts that once one has a fns weight $\phi$ ($= \psi_{\mu}$) on $M$, one can construct every normal semifinite weight $\psi$ (= $\psi_\nu$) using $\phi$. In case the density $d\nu/d\mu$ is bounded, the theorem can be reformulated thus: if $\psi \leqslant c\phi$ for some $c > 0$, there exists $h$ in $M_+$ such that $\psi(f) = \phi(hf)$ for all $f$ in $M_+$. In the case of an unbounded density, the above equality is still valid, but must be made sense of (within the language of von Neumann algebras), when $h$ is unbounded but affiliated with $M$.

We shall prove the non-commutative Radon-Nikodym theorem of Pedersen and Takesaki only in the case corresponding to finite $\mu$ and $\nu$ and bounded $d\nu/d\mu$. The result, in its full generality, will only be stated, and the reader desirous of a proof is directed to [PT].

The first problem one encounters -- even in the finite case -- is this: if $\phi \in M_{*,+}$ and if $h, x \in M_+$, there is no reason why $\phi(hx)$ should be non-negative. This is true if $\phi$ is tracial (since, then $\phi(hx) = \phi(h^{1/2}xh^{1/2}) \geqslant 0$) but not in general. Thus, the modular group $\sigma^\phi$ naturally makes its presence felt.

After this somewhat lengthy preamble, let us set the (non-commutative?) ball rolling by getting a few simple observations, in the guise of exercises, out of our way.

## Exercises

**(2.6.1)** Let $\phi$ be a faithful normal positive linear functional on $M$. If $h \in M$, let $\phi(h.)$ denote the linear functional on $M$ defined by $(\phi(h.))(x) = \phi(hx)$.

(a) The map $h \to \phi(h.)$ is a linear map from $M$ into $M_*$.
(b) If $h \in M_+^\phi$ (cf. Def. 2.5.13) and $x \in M$, then $\phi(hx) = \phi(h^{1/2}xh^{1/2})$, and consequently $\phi(h.) \in M_{*,+}$; in particular, $h,k \in M_+^\phi$ and $h \leqslant k$ imply $\phi(h.) \leqslant \phi(k.)$.
(c) If $h \in M_+^\phi$, then $\phi(h.)$ is faithful if and only if $\ker h = (0)$.
(d) If $h \in M_+^\phi$, then $\phi(h.) \circ \sigma_t^\phi = \phi(h.)$ for all $t$ in $\mathbb{R}$ (Hint: $\phi \circ \sigma_t^\phi = \phi$.). □

**Proposition 2.6.2.** *Let $\phi$ be a faithful normal positive linear functional on $M$. The following conditions on a $\psi$ in $M_{*,+}$ are equivalent:*

(i) $\psi = \phi(h.)$ *for some $h$ in* $M_+^\phi$ ;
(ii) $\psi \leqslant c\phi$ *for some* $c > 0$, *and* $\psi \circ \sigma_t^\phi = \psi$ *for all $t$ in* $\mathbb{R}$ .

**Proof.** (i) ⇒ (ii): This follows -- with $c = \|h\|$ -- from parts (b) and (d) of Ex. (2.6.1).

(ii) ⇒ (i): Assume, without loss of generality that $\phi(x) = \langle x\Omega,\Omega\rangle$ for $x$ in $M$, where $\Omega$ is a cyclic and separating vector for $M$. Notice that the sesquilinear form $[x\Omega,y\Omega] = \psi(y^*x)$ is well-defined on the dense subspace $M\Omega$ of $\mathcal{H}$, and is bounded, since

$$|\psi(y^*x)| \leqslant \psi(x^*x)^{1/2}\psi(y^*y)^{1/2} \qquad \text{(by Ex. (2.1.1) (b))}$$

$$\leqslant c\phi(x^*x)^{1/2}\ \phi(y^*y)^{1/2}$$

$$= c\|x\Omega\|\ \|y\Omega\|.$$

So there exists a bounded operator $h'$ on $\mathcal{H}$ such that $\langle h'x\Omega, y\Omega\rangle = \psi(y^*x)$ for $x$, $y$ in $M$. Note that $h' \geqslant 0$ since $\psi$ is positive; also for any $x,y,z$ in $M$,

$$\langle h'xy\Omega, z\Omega\rangle = \psi(z^*xy)$$

$$= \psi((x^*z)^*y)$$

$$= \langle h'y\Omega,\ x^*z\Omega\rangle$$

$$= \langle xh'y\Omega,\ z\Omega\rangle\ ;$$

conclude that $h' \in M'$.

Let us simply write $S,F,J$ and $\Delta$ for $S_\phi, F_\phi, J_\phi$ and $\Delta_\phi$. The $\sigma^\phi$-invariance of $\psi$ translates into the fact that $h'$ commutes with $\Delta^{it}$ for all $t$, and hence with $\Delta$. (Reason:

$$\langle h'\Delta^{it}x\Omega, y\Omega\rangle = \langle h'\sigma_t^\phi(x)\Omega,\ y\Omega\rangle = \psi(y^*\sigma_t^\phi(x)) = \psi(\sigma_{-t}^\phi(y)^*x)$$

$$= \langle h'x\Omega,\ \Delta^{-it}y\Omega\rangle = \langle \Delta^{it}h'x\Omega,\ y\Omega\rangle.)$$

Recall that $M'\Omega \subseteq \operatorname{dom} F$ and $Fa'\Omega = a'^*\Omega$ for $a'$ in $M'$; since $h' \in M'_+$, and $h'$ commutes with $\Delta$, conclude that

$$h'\Omega = Fh'\Omega = J\Delta^{-1/2}h'\Omega = Jh'\Omega.$$

It follows that $h = Jh'J \in M_+$ (by the Tomita-Takesaki Theorem) and that

$$h\Omega = Jh'J\Omega = Jh'\Omega = h'\Omega.$$

Hence, for any $x$ in $M$,

$$\psi(x) = \langle h'x\Omega, \Omega\rangle$$

$$= \langle x\Omega, h'\Omega\rangle$$

$$= \langle x\Omega, h\Omega\rangle$$

$$= \phi(hx).$$

Finally, $h'$ commutes with $\Delta^{it}$ for each $t$, as does $J$ [cf. Prop. 2.3.2 (c)] and hence,

$$h = \Delta^{it} h \Delta^{-it} = \sigma_t^{\phi}(h), \quad \text{i.e.,} \quad h \in M_+^{\phi}. \quad \square$$

With no further apology, we state below, without proof, the Radon-Nikodym theorem of Pedersen and Takesaki in its general form.

**Theorem 2.6.3.** *Let $\phi$ be a* fns *weight on $M$. Let $\psi$ be a normal semifinite weight on $M$ such that $\psi \circ \sigma_t^{\phi} = \psi$ for all $t$. Then there exists a unique positive self-adjoint operator $H$ (possibly unbounded) affiliated to $M^{\phi}$ such that $\psi = \phi(H.)$; where $\phi(H.)$ is defined to be the limit (as $\epsilon \to 0$) of the increasing net $\{\phi(H_{\epsilon}.): \epsilon > 0\}$ (directed so that $\epsilon_1 < \epsilon_2 \Rightarrow \phi(H_{\epsilon_1}.) \geqslant \phi(H_{\epsilon_2}.)$) of normal semifinite weights on $M$ defined by $(\phi(H_{\epsilon}.))(x) = \phi(H_{\epsilon}^{1/2} x H_{\epsilon}^{1/2})$,* where *$H_{\epsilon} = H(1 + \epsilon H)^{-1}$.* $\square$

As might be expected, this result will suffer the same fate as other unproved results concerning semifinite weights: we shall use it in the future with complete equanimity.

The rest of this section is a digression, as far as the subsequent trend of this book is concerned. The reader who has had no prior exposure to probability theory, who might consequently not appreciate the rest of this section may safely proceed to the next chapter.

Let $M_0$ be a von Neumann subalgebra of $M$. If $M = L^{\infty}(X,\mathcal{F},\mu)$, it follows, from the fact that $M_0$ is generated by its projections, that $M_0 = L^{\infty}(X,\mathcal{F}_0,\mu)$ where $\mathcal{F}_0$ (is the $\sigma$-subalgebra of $\mathcal{F}$ which) consists of those sets in $\mathcal{F}$, multiplication by whose indicator function defines a projection in $M_0$. When $\mu$ is a probability measure (i.e., $\mu(X) = 1$) the classical conditional expectation is a linear map $E: M \to M_0$ satisfying: (i) $x \geqslant 0$ implies $Ex \geqslant 0$; (ii) $E$ is a projection of norm one; (iii) $E$ is normal, in that it respects monotone limits; and (iv) $\phi \circ E = \phi$, where $\phi$ is the faithful normal state defined by $\phi(f) = \int f \, d\mu$ for $f$ in $M$. Notice that $\phi$ is a faithful normal state on $M$, so that $\phi_0 = \phi|M_0$ is a faithful normal state on $M_0$; the GNS triples for $(M_0,\phi_0)$ and $(M,\phi)$ are $(L^2(X,\mathcal{F}_0,\mu),m.,\Omega)$ and $(L^2(X,\mathcal{F},\mu),m.,\Omega)$, where $m.$ is the representation $f \to m_f$ and $\Omega$ is the constant function 1. The GNS space $\mathcal{H}_0$ for $(M_0,\phi_0)$ sits naturally as a subspace of the GNS space $\mathcal{H}$ for $(M,\phi)$, and it is well-known (and easy to derive, from the properties (i) - (iv) listed above) that

$$p_{\mathcal{H}_0}(\pi_{\phi}(x)\Omega) = \pi_{\phi_0}(Ex)\Omega .$$

We shall commence the non-commutative proceedings with an old result due to Tomiyama on norm one projections. The result is valid in the context of $C^*$-algebras and may be inferred from the version given below, via the so-called enveloping von Neumann algebra; we shall, however, be content with the result for von Neumann algebras.

**Proposition 2.6.4.** *Let $M_0$ be a von Neumann subalgebra of $M$. If $E$: $M \to M_0$ is a projection of norm one, which is normal, then*

(a) $E \geqslant 0$; *i.e.*, $x \in M_+ \Rightarrow Ex \in M_{0,+}$;
(b) $E(a_0xb_0) = a_0(Ex)b_0$, *if* $x \in M$, $a_0,b_0 \in M_0$;
*and*
(c) $(Ex)^*(Ex) \leqslant E(x^*x)$, *for all* $x$ *in* $M$.

**Proof.** (a) It suffices to prove that if $x \in M_+$ and $\phi_0 \in M_{0,+}^*$, then $\phi_0(Ex) \geqslant 0$; in other words, we must show that $\phi_0 \in M_{0,+}^* \Rightarrow \phi_0 \circ E \in M_+^*$. For this, note that

$$\phi_0(E1) = \phi_0(1) = \|\phi_0\| \geqslant \|\phi_0 \circ E\| \geqslant |\phi_0(E1)|,$$

so that $\phi_0 \circ E$ attains its norm at the identity element. This implies that $\phi_0 \circ E \in M_+^*$ (cf. the parenthetical remark in Ex. (2.1.1) (c)).

(b) Since $E$ is normal and preserves adjoints (thanks to (a)), it suffices to show that $E(e_0x) = e_0E(x)$ whenever $e_0 \in \mathcal{P}(M_0)$ and $0 \leqslant x \leqslant 1$. Let $f_1 = e_0$ and $f_2 = e_0^\perp = 1 - e_0$.

(1) Assertion: $E(f_ixf_j) = f_iE(f_ixf_j)f_j, \quad 1 \leqslant i,j \leqslant 2.$

For $f = f_i$, $i = 1,2$, notice that $0 \leqslant fxf \leqslant f$ since $0 \leqslant x \leqslant 1$; hence, by (a), $0 \leqslant E(fxf) \leqslant Ef = f$, since $f \in M_0$. This implies that $E(fxf) = fE(fxf)f$, thereby establishing (1) when $i = j$. For $i \neq j$, it suffices (by considering adjoints) to consider the case $i = 1$, $j = 2$; thus we must show that if $x_0 = E(e_0xe_0^\perp)$, then

$$e_0x_0e_0 = e_0^\perp x_0e_0^\perp = e_0^\perp x_0e_0 = 0.$$

Notice that for any $\lambda$ in $\mathbb{C}$,

$$\begin{aligned}\|x_0 + \lambda e_0\|^2 &= \|E(e_0xe_0^\perp + \lambda e_0)\|^2 \\ &\leqslant \|e_0xe_0^\perp + \lambda e_0\|^2 \\ &= \|(e_0xe_0^\perp + \lambda e_0)(e_0xe_0^\perp + \lambda e_0)^*\| \\ &= \|e_0xe_0^\perp xe_0 + |\lambda|^2e_0\| \\ &\leqslant 1 + |\lambda|^2.\end{aligned} \tag{2}$$

On the other hand, if $\alpha \in \mathrm{sp}(\mathrm{Re}(e_0x_0e_0))$ -- where $\mathrm{Re}\, y = \frac{1}{2}(y + y^*)$ -- and if $\lambda$ is any real number such that $\alpha\lambda \geqslant 0$, then

$$\begin{aligned}\|x_0 + \lambda e_0\| &\geqslant \|e_0(x_0 + \lambda e_0)e_0\| \\ &\geqslant \|\mathrm{Re}(e_0x_0e_0 + \lambda e_0)\| \geqslant |\alpha + \lambda|.\end{aligned} \tag{3}$$

If (2) and (3) are to be compatible for all $\lambda$ of the same sign as $\alpha$ and of arbitrarily large modulus, it must be the case that $\alpha = 0$. Since $\mathrm{Re}(e_0x_0e_0)$ is self-adjoint and $\alpha$ was an arbitrary number in its spectrum, conclude that $\mathrm{Re}(e_0x_0e_0) = 0$. An exactly similar reasoning shows that $\mathrm{Im}(e_0x_0e_0) = 0$, whence $e_0x_0e_0 = 0$.

Reversing the roles of $e_0$ and $e_0^\perp$ in the above reasoning, we find that $e_0^\perp E(e_0^\perp xe_0)e_0^\perp = 0$; take adjoints to conclude that $e_0^\perp x_0e_0^\perp = 0$.

The conclusions of the preceding paragraphs show that, with respect to the decomposition $\mathcal{H} = e_0\mathcal{H} \oplus e_0^\perp\mathcal{H}$, the operator $x_0$ is represented by a matrix of the form

$$\begin{bmatrix} 0 & a \\ b & 0 \end{bmatrix}.$$

To complete the proof of the assertion, we must show that $b = 0$.

Minor computations reveal that for any scalar $\lambda$,

$$\begin{bmatrix} 0 & a \\ (\lambda+1)b & 0 \end{bmatrix} = x_0 + \lambda e_0^\perp x_0e_0 = E(e_0xe_0^\perp + \lambda e_0^\perp x_0e_0),$$

and hence,

$$\begin{aligned} \max\{|\lambda+1|\ \|b\|, \|a\|\} &= \|E(e_0xe_0^\perp + \lambda e_0^\perp x_0e_0)\| \\ &\leqslant \|e_0xe_0^\perp + \lambda e_0^\perp x_0e_0\| \\ &= \left\| \begin{bmatrix} 0 & c \\ \lambda b & 0 \end{bmatrix} \right\| \\ &= \max\{|\lambda|\ \|b\|, \|c\|\}, \end{aligned}$$

where $c = (e_0xe_0^\perp \mid e_0^\perp\mathcal{H})\colon e_0^\perp\mathcal{H} \to e_0\mathcal{H}$. The validity of this inequality for large positive $\lambda$ forces $\|b\| = 0$, and the assertion is proved.

Conclude, finally, that

$$\begin{aligned} e_0E(x) &= f_1E\left[\sum_{i,j=1}^{2} f_ixf_j\right] \\ &= \sum_{j=1}^{2} f_1E(f_1xf_j)f_j \\ &= \sum_{j=1}^{2} E(f_1xf_j) \\ &= E(f_1x) = E(e_0x), \end{aligned}$$

and (b) is proved.

(c) $$0 \leq E((x - Ex)^*(x - Ex))$$
$$= E(x^*x - (Ex)^*x - x^*Ex + (Ex)^*(Ex))$$
$$= E(x^*x) - (Ex)^*(Ex), \quad \text{by (b).} \quad \square$$

**Definition 2.6.5.** Let $M_0$ be a von Neumann subalgebra of $M$.

(a) A normal projection of norm one of $M$ onto $M_0$ will be called a conditional expectation of $M$ onto $M_0$.
(b) If $\phi$ is a faithful normal state on $M$, a conditional expectation $E$ of $M$ onto $M_0$ is said to be $\phi$-compatible if $\phi \circ E = \phi$. □

Hence, Tomiyama's result lists some properties of a general conditional expectation, which -- particularly (b) of Prop. 2.6.4 -- will justify the use of the term "conditional expectation" in the eyes of a probabilist. To a probabilist, however, the notion that we have called "$\phi$-compatibility" is the crux of the matter.

When $M = L^\infty(X,\mathcal{F},\mu)$, the classical conditional expectation settles the question of the existence of $\phi$-compatible conditional expectations. For a general non-abelian $M$, the modular group $\sigma^\phi$ intervenes as an obstruction; the following result shows that it is the only obstruction.

**Proposition 2.6.6.** *Let $M_0$ be a von Neumann subalgebra of $M$, and let $\phi$ be a faithful normal state on $M$. The following conditions are equivalent*:

(i) *there exists a $\phi$-compatible conditional expectation $E$ of $M$ onto $M_0$*;

(ii) $\sigma_t^\phi(M_0) \subseteq M_0$ *for all $t$ in $\mathbb{R}$*;

(iii) $\sigma_t^\phi(x_0) = \sigma_t^{\phi_0}(x_0)$ *for all $x_0$ in $M_0$ and $t$ in $\mathbb{R}$, where $\sigma^{\phi_0}$ is the modular group of automorphisms of $M_0$ corresponding to the faithful normal state $\phi_0$ on $M_0$ given by $\phi_0 = \phi|M_0$.*

**Proof.** It is clear that $\phi_0 = \phi|M_0$ is a faithful normal state on $M_0$. Let $(\mathcal{H},\pi,\Omega)$ be the GNS triple for $(M,\phi)$, and let $\mathcal{H}_0 = \overline{\pi(M_0)\Omega}$. Since $\mathcal{H}_0$ is invariant under the self-adjoint algebra $\pi(M_0)$, so is $\mathcal{H}_1 = \mathcal{H} \ominus \mathcal{H}_0$; consequently, each operator in $\pi(M_0)$ is of the form $\pi_0(x_0) \oplus \pi_1(x_0)$ (for some $x_0$ in $M_0$) with respect to the decomposition $\mathcal{H} = \mathcal{H}_0 \oplus \mathcal{H}_1$, where $\pi_j$ is a normal *-homomorphism of $M_0$ into $\mathcal{L}(\mathcal{H}_j)$ for $j = 0,1$. It is trivial to verify that $(\mathcal{H}_0,\pi_0,\Omega)$ is a GNS triple for $(M_0,\phi_0)$. Let $S,F,J$ and $\Delta$ (resp., $S_0$, $F_0$, $J_0$ and $\Delta_0$) denote the "modular operators" on $\mathcal{H}$ (resp., $\mathcal{H}_0$) associated with $(M,\phi)$ (resp., $(M_0,\phi_0)$) via the Tomita-Takesaki construction.

**(i) ⇒ (ii).** If $x \in M$ and $x_0 \in M_0$, then

$$\begin{aligned} <\pi(x)\Omega, \pi(x_0)\Omega> &= \phi(x_0^* x) \\ &= \phi(E(x_0^* x)) \\ &= \phi(x_0^* E(x)) \\ &= <\pi(E(x))\Omega, \pi(x_0)\Omega> ; \end{aligned}$$

since $\overline{\pi(M_0)\Omega} = \mathcal{H}_0$, conclude that if $p = p_{\mathcal{H}_0}$, then $p\pi(x)\Omega = \pi(Ex)\Omega$.

Consequently $p(\pi(M)\Omega) \subseteq \text{dom } S$; also, for any $x$ in $M$, $Sp(\pi(x)\Omega) = \pi((Ex)^*)\Omega = \pi(Ex^*)\Omega = pS(\pi(x)\Omega)$. Since $\pi(M)\Omega$ is a core for $S$ (by definition of $S$), this implies that $pS \subseteq Sp$. It also follows from the above equation that $S_0 = S|(\text{dom } S \cap \mathcal{H}_0)$ and that in fact $S = S_0 \oplus S_1$ (for an appropriate conjugate linear closed operator $S_1$ in $\mathcal{H}_1$) with respect to the decomposition $\mathcal{H} = \mathcal{H}_0 \oplus \mathcal{H}_1$ -- the direct sum of unbounded operators being defined in the natural way (cf. Ex. (2.5.6)). (In case the reader feels he is being hoodwinked by a case of somewhat excessive "hand-waving", he may be pleased to know that the gruesome details of the verification of the preceding statements are spelt out in Ex. (2.6.7).)

It follows easily now that all the "modular operators" admit direct sum decompositions: $F = F_0 \oplus F_1$, $J = J_0 \oplus J_1$ and $\Delta = \Delta_0 \oplus \Delta_1$. In particular, if $x_0 \in M_0$, $t \in \mathbb{R}$,

$$\begin{aligned} \pi(\sigma_t^\phi(x_0))\Omega &= \Delta^{it}\, \pi(x_0)\Delta^{-it}\Omega \\ &= \Delta_0^{it}\pi_0(x_0)\Delta_0^{-it}\Omega \qquad (\text{since } \Omega \in \mathcal{H}_0) \\ &= \pi_0(\sigma_t^{\phi_0}(x_0))\Omega = \pi(\sigma_t^{\phi_0}(x_0))\Omega . \end{aligned}$$

Since $\Omega$ is a separating vector for $\pi_\phi(M)$, conclude that

$$\sigma_t^\phi(x_0) = \sigma_t^{\phi_0}(x_0) \in M_0 .$$

(We have actually proved (i) ⇒ (iii), but clearly (iii) ⇒ (ii).)

**(ii) ⇒ (iii).** If $\sigma_t^\phi(M_0) \subseteq M_0$, then $M_0 \subseteq \sigma_{-t}^\phi(M_0)$, and so, the assumption is that $\sigma_t^\phi(M_0) = M_0$ for all $t$. The equation $\alpha_t = \sigma_t^\phi|M_0$ clearly defines a flow on $M_0$ with respect to which $\phi_0$ satisfies the KMS condition (since $\phi$ satisfies the KMS condition with respect to $\sigma_t^\phi$); hence, by Theorem 2.5.11, $\alpha_t = \sigma_t^{\phi_0}$ for all $t$.

**(iii) ⇒ (i).** Observe, to begin with, that

$$\begin{aligned}
\Delta_0^{it}(\pi(x_0)\Omega) &= \Delta_0^{it}(\pi_0(x_0)\Omega) \\
&= \pi_0(\sigma_t^{\phi_0}(x_0))\Omega \qquad \text{(by Ex. (2.3.6) (b))} \\
&= \pi(\sigma_t^{\phi_0}(x_0))\Omega \\
&= \pi(\sigma_t^{\phi}(x_0))\Omega \qquad \text{(by hypothesis (iii))} \\
&= \Delta^{it}(\pi(x_0)\Omega) \qquad \text{(again by Ex. (2.3.6) (b)).}
\end{aligned}$$

Since $\overline{\pi(M_0)\Omega} = \mathcal{H}_0$, this implies that $\Delta^{it}p = p\Delta^{it}$ and that $\Delta^{it}|\mathcal{H}_0 = \Delta_0^{it}$. Since $t$ is arbitrary, this forces $p\Delta \subseteq \Delta p$ and $\Delta_0 = \Delta|(\text{dom } \Delta \cap \mathcal{H}_0)$. Further,

$$\begin{aligned}
J(\Delta^{1/2}\pi(x_0)\Omega) &= \pi(x_0^*)\Omega = \pi_0(x_0^*)\Omega = J_0\Delta_0^{1/2}\pi_0(x_0)\Omega \\
&= J_0(\Delta^{1/2}\pi(x_0)\Omega).
\end{aligned}$$

Note that $\pi(M_0)\Omega$ is a core for $\Delta_0^{1/2}$, which is invertible, and so, $\Delta_0^{1/2}\pi_0(M_0)\Omega$ $(= \Delta^{1/2}\pi(M_0)\Omega)$ is dense in $\mathcal{H}_0$. Conclude from the preceding equality that $J\mathcal{H}_0 \subseteq \mathcal{H}_0$ and $J|\mathcal{H}_0 = J_0$. Since $J$ is self-adjoint, it follows that $J\mathcal{H}_0 = \mathcal{H}_0$ and $J = J_0 \oplus J_1$, where $J_1$ is a self-adjoint antiunitary operator on $\mathcal{H}_1$.

**Assertion.** If $x \in M$ and $\pi(x)$ has the operator matrix

$$\begin{bmatrix} x_{11} & x_{12} \\ x_{21} & x_{22} \end{bmatrix}$$

with respect to the decomposition $\mathcal{H} = \mathcal{H}_0 \oplus \mathcal{H}_1$, then $x_{11} \in \pi_0(M_0)$.

If $a_0' \in \pi_0(M_0)'$, then $J_0a_0'J_0 \in \pi_0(M_0)$; if $x_0 \in M_0$ and $J_0a_0'J_0 = \pi_0(x_0)$, then $\pi(x_0) = \pi_0(x_0) \oplus \pi_1(x_0) = J_0a_0'J_0 \oplus \pi_1(x_0)$. Note that $J\pi(x_0)J \in \pi(M)'$, whereas $J\pi(x_0)J = a_0' \oplus J_1\pi_1(x_0)J_1$, since $J = J_0 \oplus J_1$. Hence $\pi(x)$ commutes with $a_0' \oplus J_1\pi_1(x_0)J_1$. Comparing the (1,1)-entries of the two products, we find that $x_{11}a_0' = a_0'x_{11}$. The arbitrariness of $a_0'$ and the double commutant theorem now settle the assertion.

To complete the proof of (iii) $\Rightarrow$ (i), simply define $E: M \to M_0$ by $E(x) = \pi_0^{-1}(x_{11})$, with $x$ and $x_{11}$ related as in the assertion. It is an easy matter to verify now that $E$ is a $\phi$-compatible conditional expectation.

## Exercises

**(2.6.7)** Let $A$ be a closed densely defined operator in $\mathcal{H} = \mathcal{H}_1 \oplus \mathcal{H}_2$, with dom $A = \mathcal{D}$. Let $p_i = p_{\mathcal{H}_i}$ for $i = 1,2$. Let $A^0 = A|\mathcal{D}^0$, where $\mathcal{D}^0$ is a core for $A$. (In this exercise, either all the operators are linear, or all the operators, but for $p_1$ and $p_2$, are conjugate linear.) Suppose that $p_1A^0 \subseteq A^0p_1$.

(a) $p_2A^0 \subseteq A^0p_2$, and in particular, $p_i\mathfrak{D}^0 \subseteq \mathfrak{D}^0$, $i = 1,2$. (Hint: $\mathfrak{D}^0$ is a linear subspace and $p_2\xi = \xi - p_1\xi$.)

(b) If $\mathfrak{D}_i^0 = \mathfrak{D}^0 \cap \mathcal{H}_i$, then $\mathfrak{D}_i^0 = p_i\mathfrak{D}^0$ and hence $\mathfrak{D}_i^0$ is dense in $\mathcal{H}_i$, for $i = 1,2$. (Hint: $\mathfrak{D}_i^0 = p_i(\mathfrak{D}_i^0) \subseteq p_i(\mathfrak{D}^0) \cap \mathcal{H}_i \subseteq \mathfrak{D}^0 \cap \mathcal{H}_i$, while $\mathfrak{D}^0$ being a core for the densely defined $A$ is necessarily dense in $\mathcal{H}$.)

(c) $p_iA \subseteq Ap_i$ and $p_i(\mathfrak{D}) = \mathfrak{D} \cap \mathcal{H}_i$, for $i = 1,2$. (Hint: by definition of $\mathfrak{D}^0$, if $\xi \in \mathfrak{D}$, then there exist $\xi_n$ in $\mathfrak{D}^0$ such that $\xi_n \to \xi$, $A_0\xi_n \to A\xi$.)

(d) Let $\mathfrak{D}_i = p_i(\mathfrak{D})$. Since $A(\mathfrak{D}_i) \subseteq \mathcal{H}_i$, define an operator $A_i$ in $\mathcal{H}_i$ with dom $A_i = \mathfrak{D}_i$, by $A_i\xi = A\xi$ for $\xi$ in $\mathfrak{D}_i$. Show that $A_i$ is closed and that $\mathfrak{D}_i^0$ is a core for $A_i$, for $i = 1,2$. (Hint: Identify $\mathcal{H}_i$ with a closed subspace of $\mathcal{H}$ and notice that $G(A_i) = G(A) \cap (\mathcal{H}_i \oplus \mathcal{H}_i)$.)

(e) Show that $A = A_1 \oplus A_2$ in the sense that dom $A = \{\xi_1 \oplus \xi_2 \in \mathcal{H}_1 \oplus \mathcal{H}_2 : \xi_i \in \text{dom } A_i, i = 1,2\}$ and $A(\xi_1 \oplus \xi_2) = (A_1\xi_1) \oplus (A_2\xi_2)$. □

**Example 2.6.8.** Let $M = \mathcal{L}(\mathcal{H})$ with dim $\mathcal{H} > 2$. Let $e$ be a projection of rank one, and let $M_0 = \{e\}'$. Let $\rho$ be a positive invertible trace-class operator with tr $\rho = 1$. Then $\phi(x) = \text{tr } \rho x$ defines a faithful normal state on $M$. Recall -- from Example 2.3.7 (b) -- that $\sigma_t^\phi(x) = \rho^{it}x\rho^{-it}$ for $x \in M$, $t \in \mathbb{R}$. Notice that $\rho^{it}M_0\rho^{-it} = M_0 \Leftrightarrow \rho^{it}M_0'\rho^{-it} = M_0'$; however, (by the remarks following the double commutant theorem) $M_0' = \{e\}'' = \{\lambda e + \mu : \lambda,\mu \in \mathbb{C}\}$. Notice that $\mathcal{P}(M_0') = \{0,e,1-e,1\}$ and that $e$ is the only rank one projection in $M_0'$. Hence, $\rho^{it}M_0\rho^{-it} = M_0 \Leftrightarrow \rho^{it}e\rho^{-it} = e$. Thus, there exists a $\phi$-compatible conditional expectation of $M$ onto $M \Leftrightarrow \rho^{it}e = e\rho^{it}$ $\forall t$ in $\mathbb{R} \Leftrightarrow \rho e = e\rho$. Thus, even for finite-dimensional von Neumann algebras, $\phi$-compatible conditional expectations need not exist. □

**Remarks 2.6.9.** (a) There is an extension of Prop. 2.6.6 to the case when $\phi$ is a fns weight. The result -- which is the theme of [Tak 2] -- is that if $\phi$ is a fns weight on $M$ such that $\phi|M_{0,+}$ is still a semifinite weight, then the conditions (i) - (iii) of Prop. 2.6.6 are equivalent. (Recall the $\sigma$-finiteness hypothesis in the classical Radon- Nikodym theorem, without which hypothesis, that theorem is false.)

(b) Even when the condition (iii) of Prop. 2.6.6 is not satisfied, it is possible to pursue the reasoning of the proof of (iii) ⇒ (i) as follows: with the notation of the proposition, notice that if $x \in M$, then $J\pi(x)J \in \pi(M)' \subseteq \pi(M_0)'$. If

$$J\pi(x)J = \begin{bmatrix} a'_{11} & a'_{12} \\ a'_{21} & a'_{22} \end{bmatrix},$$

in the decomposition $\mathcal{H} = \mathcal{H}_0 \oplus \mathcal{H}_1$, it follows -- as in the proof of the Assertion -- that $a'_{11} \in \pi_0(M_0)'$, and hence that $J_0 a'_{11} J_0 \in \pi_0(M_0)$. Define $E$: $M \to M_0$ by $E(x) = \pi_0^{-1}(J_0 a'_{11} J_0)$. It is clear that $E$ is a positivity-preserving, normal, contractive linear operator from $M$ to $M_0$. (For the reader to whom this makes sense, $E$ is also completely positive, being a composite of normal homomorphisms and a "compression-operation".) It is not difficult to show that this $E$ is a projection onto $M_0$ -- i.e., fixes elements of $M_0$ -- precisely when the condition (iii) is met, in which case it is the unique $\phi$-compatible conditional expectation of $M$ onto $M_0$. In the general case, the map $E$ defined as above induces a partial isometry -- not necessarily a projection -- on the GNS space $\mathcal{H}_\phi$, which sends $\pi_\phi(x)\Omega_\phi$ to $\pi_\phi(Ex)\Omega_\phi$. It can even turn out that the final space of this partial isometry is a proper subspace of $\mathcal{H}_0$. It may even be the case that $E(M)$ is not a von Neumann algebra of operators on $\mathcal{H}_0$. For details of such an investigation, the reader may consult [AC]. □

# Chapter 3
# THE CONNES CLASSIFICATION OF TYPE III FACTORS

The first section discusses the extent to which the modular group $\sigma^{\phi}$ depends upon the fns weight $\phi$. The precise description is the unitary cocycle theorem of Connes, which says, loosely, that modulo the group of inner automorphisms of $M$, the modular group $\sigma^{\phi}$ is independent of $\phi$.

Stone's theorem states that every strongly continuous unitary representation $t \to u_t$ of the real line $\mathbb{R}$ in a Hilbert space $\mathcal{H}$ is given by $u_t = e^{itH}$ for a uniquely determined self-adjoint operator $H$ in $\mathcal{H}$. Taking a cue from the physicists, one may regard sp $H$ as "the spectrum of the representation $\{u_t\}$". Arveson's idea is to imitate this procedure for flows on a von Neumann algebra. Since the proofs are no harder in the more general setting of locally compact abelian groups (rather than just $\mathbb{R}$), the general case is treated in Section 3.2, which begins with a rapid survey of the necessary results from abstract harmonic analysis, and goes on to the definition and some elementary propositions concerning the Arveson spectrum of a group action on a von Neumann algebra, the said "spectrum" being a certain closed subset of the dual group.

The Arveson spectrum of the modular group $\{\sigma_t^{\phi}\}$ would, in general, vary with the fns weight $\phi$; Section 3.3 introduces the Connes spectrum of a group action, which is a refinement of the Arveson spectrum and has the following pleasing properties: (a) the Connes spectrum of an action of $G$ on $M$ is a closed subgroup of the dual group $\Gamma$; and (b) if $\phi$ and $\psi$ are any two fns weights on $M$, the Connes spectra of $\{\sigma_t^{\phi}\}$ and $\{\sigma_t^{\psi}\}$ coincide. Thus one may define $\Gamma(M)$ to be the Connes spectrum of $\{\sigma_t^{\phi}\}$ where $\phi$ is any fns weight on $M$. Since the closed subgroups of $\mathbb{R}$ are easily enumerated, the invariant $\Gamma(M)$ leads to a refinement of the Murray-von Neumann classification.

The definition of $\Gamma(M)$ given in Section 3.3 is somewhat unmanageable, for computational purposes; Section 3.4 is devoted to establishing other descriptions of $\Gamma(M)$. One of these descriptions is in terms of another invariant $S(M)$ which, in turn, can be described in terms of the modular operator $\Delta_\phi$ associated with one given fns weight on $M$; it is this description which will become useful in Section 4.3, which is devoted to the construction of examples of factors of the various types.

## 3.1. The Unitary Cocycle Theorem

If $M$ is a von Neumann algebra, the symbol $U(M)$ will denote, in the sequel, the group of unitary operators in $M$. It is a basic fact that, restricted to $U(M)$, the weak and strong topologies coincide. (Reason: if $u_i \to u$ weakly, and if $\xi \in \mathcal{H}$, then $\|(u_i - u)\xi\|^2 = 2\|\xi\|^2 - 2 \operatorname{Re} \langle u_i\xi, u\xi\rangle \to 0$.) With this bit of trivia out of the way, let us proceed to the heart of this section.

**Theorem 3.1.1.** *If $\phi$ and $\psi$ are* fns *weights on $M$, there exists a strongly continuous map $t \to u_t$ from $\mathbb{R}$ into $U(M)$ such that*

(a) $\sigma_t^\psi(x) = u_t\, \sigma_t^\phi(x) u_t^*$ *for all $x$ in $M$, $t$ in $\mathbb{R}$,*
*and*
(b) $u_{t+s} = u_t \sigma_t^\phi(u_s)$, *for all $s,t$ in $\mathbb{R}$.*

**Proof.** Let $\tilde{\mathcal{H}} = \mathcal{H} \oplus \mathcal{H}$; we shall identify operators $\tilde{x}$ on $\tilde{\mathcal{H}}$ with $2 \times 2$ matrices $((x_{ij}))$, where $x_{ij} \in \mathcal{L}(\mathcal{H})$ for $1 \leq i,j \leq 2$. There is a natural identification $\mathcal{L}(\tilde{\mathcal{H}}) \cong \mathcal{L}(\mathcal{H}) \otimes M_2(\mathbb{C})$ (where $M_2(\mathbb{C})$ is the set of $2 \times 2$ complex matrices) whereby

$$\tilde{x} \leftrightarrow \sum_{i,j=1}^{2} x_{ij} \otimes e_{ij}\,,$$

where $e_{ij}$ is the $2 \times 2$ matrix with 1 in the $(i,j)$ place and 0 elsewhere.

Let

$$\tilde{M} = M \otimes M_2(\mathbb{C}) = \left\{ \begin{bmatrix} x_{11} & x_{12} \\ x_{21} & x_{22} \end{bmatrix} \in \mathcal{L}(\tilde{\mathcal{H}}) : x_{ij} \in M \right\}.$$

It is readily verified that $\tilde{M}$ is a von Neumann algebra of operators on $\tilde{\mathcal{H}}$. Notice that $\tilde{x} \in \tilde{M}_+ \Rightarrow x_{11}, x_{22} \in M_+$ and that when $\tilde{x} \geq 0$, $x_{11} = x_{22} = 0 \Rightarrow \tilde{x} = 0$. (Check this, by examining $\langle \tilde{x}\xi,\xi\rangle$ where $\xi = \lambda_1\xi_1 \oplus \lambda_2\xi_2$ for arbitrary $\lambda_i \in \mathbb{C}$ and temporarily fixed $\xi_i$ in $\mathcal{H}$.) Consequently the equation $\theta(\tilde{x}) = \phi(x_{11}) + \psi(x_{22})$, $\tilde{x} \in \tilde{M}_+$, is meaningful and is easily seen to define a faithful, normal weight. Furthermore $\theta$ is semifinite. (Reason: since $\phi$ (resp., $\psi$) is semifinite, there exists (cf. Ex. (2.4.8)) a monotone net $\{x_i : i \in I\} \subseteq \mathfrak{D}_\phi$ (resp., $\{y_j : j \in J\} \subseteq \mathfrak{D}_\psi$) such that $x_i \nearrow 1$ (resp., $y_j \nearrow 1$). The set $K = I \times J$ is directed upwards with respect to the order $(i_1,j_1) \leq (i_2,j_2) \Leftrightarrow i_1 \leq i_2$

and $j_1 \leqslant j_2$; the net $\{x_i \oplus y_j: (i,j) \in K\}$ is monotone, lies in $\mathfrak{D}_\theta$ and $x_i \oplus y_j \nearrow 1_{\tilde{\mathcal{H}}}$.)

Since

$$(\widetilde{x}^*\widetilde{x})_{j,j} = \sum_{i=1}^{2} x_{ij}^* x_{ij},$$

conclude that $\widetilde{x} \in N_\theta \Leftrightarrow x_{11}, x_{21} \in N_\phi$ and $x_{12}, x_{22} \in N_\psi$. Thus, if $\widetilde{x} \in N_\theta$ and if $\widetilde{y}$ is the matrix obtained by setting some of the matrix entries of $\widetilde{x}$ equal to zero and leaving the other entries unchanged, then $\widetilde{y} \in N_\theta$; in particular, $N_\theta(1 \otimes e_{11}) \subseteq N_\theta$, and so $M_\theta(1 \otimes e_{11}) \subseteq M_\theta$. Since $M_\theta$ is self-adjoint, as is $1 \otimes e_{11}$, this implies that also $(1 \otimes e_{11})M_\theta \subseteq M_\theta$. Finally, some simple matrix multiplication shows that if $\widetilde{x},\widetilde{y} \in N_\theta$,

$$\theta((1 \otimes e_{11})\widetilde{x}^*\widetilde{y}) = \phi(x_{11}^* y_{11} + x_{21}^* y_{21})$$

$$= \theta(\widetilde{x}^*\widetilde{y}(1 \otimes e_{11})).$$

Conclude, by Theorem 2.5.14' that $(1 \otimes e_{11}) \in \widetilde{M}^\theta$; so, for any $x$ in $M$ and $t$ in $\mathbb{R}$,

$$\sigma_t^\theta(x \otimes e_{11}) = \sigma_t^\theta((1 \otimes e_{11})(x \otimes e_{11})(1 \otimes e_{11}))$$

$$= (1 \otimes e_{11})(\sigma_t^\theta(x \otimes e_{11}))(1 \otimes e_{11});$$

hence $\sigma_t^\theta(x \otimes e_{11}) = \alpha_t(x) \otimes e_{11}$ for some $\alpha_t(x) \in M$. A routine verification shows that $\{\alpha_t\}$ is a flow on $M$, in the sense of Def. 2.5.7.

**Assertion:** $\phi$ satisfies the KMS condition with respect to $\{\alpha_t\}$ and hence $\alpha_t = \sigma_t^\phi \ \forall t$.

First, if $x \in M_+$,

$$\phi(\alpha_t(x)) = \theta(\alpha_t(x) \otimes e_{11}) = \theta(\sigma_t^\theta(x \otimes e_{11})) = \theta(x \otimes e_{11}) = \phi(x).$$

Next, if $x,y \in N_\phi \cap N_\phi^*$, notice that $x \otimes e_{11}, y \otimes e_{11} \in N_\theta \cap N_\theta^*$, and that

$$\theta((\sigma_t^\theta(y \otimes e_{11}))(x \otimes e_{11})) = \phi(\alpha_t(y)x)$$

and

$$\theta((x \otimes e_{11})(\sigma_t^\phi(y \otimes e_{11})) = \phi(x\alpha_t(y));$$

thus if $F$ is KMS-admissible for $x \otimes e_{11}$ and $y \otimes e_{11}$ (relative to $\theta$), then $F$ is KMS-admissible for $x$ and $y$ (relative to $\phi$), and the assertion is proved.

We have shown that $(1 \otimes e_{11}) \in \widetilde{M}^\theta$ and that $\sigma_t^\theta(x \otimes e_{11}) = \sigma_t^\phi(x) \otimes e_{11}$ for $x \in M$, $t \in \mathbb{R}$. In an entirely analogous manner, it may be seen that $(1 \otimes e_{22}) \in \widetilde{M}^\theta$ and that $\sigma_t^\theta(x \otimes e_{22}) = \sigma_t^\psi(x) \otimes e_{22}$ for $x \in M$, $t \in \mathbb{R}$.

Since $e_{21} = e_{22}e_{21}e_{11}$, conclude that $\sigma_t^\theta(1 \otimes e_{21}) = (1 \otimes e_{22})(\sigma_t^\theta(1 \otimes

$e_{21}))(1 \otimes e_{11})$, and hence that $\sigma_t^\theta(1 \otimes e_{21}) = u_t \otimes e_{21}$ for some $u_t \in M$. Apply $\sigma_t^\theta$ to the equations $(1 \otimes e_{21})^*(1 \otimes e_{21}) = 1 \otimes e_{11}$ and $(1 \otimes e_{21})(1 \otimes e_{21})^* = 1 \otimes e_{22}$, bear in mind that $1 \otimes e_{11}$ and $1 \otimes e_{22}$ are fixed by $\sigma_t^\theta$, and find that $u_t^* u_t = u_t u_t^* = 1$; i.e., $u_t \in \mathcal{U}(M)$. Since $t \to \sigma_t^\theta(1 \otimes e_{21})$ is $\sigma$-weakly (and hence weakly) continuous, infer from the initial paragraph of this section that $t \to u_t$ is a strongly continuous map.

Apply $\sigma_t^\theta$ to the equation $(x \otimes e_{22}) = (1 \otimes e_{21})(x \otimes e_{11})(1 \otimes e_{12})$, to conclude that $\sigma_t^\psi(x) = u_t \sigma_t^\phi(x) u_t^*$. Finally, if $s,t \in \mathbb{R}$,

$$\begin{aligned} u_{t+s} \otimes e_{21} &= \sigma_t^\theta \circ \sigma_s^\theta(1 \otimes e_{21}) \\ &= \sigma_t^\theta(u_s \otimes e_{21}) \\ &= \sigma_t^\theta((1 \otimes e_{21})(u_s \otimes e_{11})) \\ &= u_t \sigma_t^\phi(u_s) \otimes e_{21}, \end{aligned}$$

thereby establishing the theorem. □

**Definition 3.1.2.** (a) An automorphism $\alpha$ of $M$ is called inner if there exists $u$ in $\mathcal{U}(M)$ such that $\alpha(x) = uxu^*$ for all $x$ in $M$.

(b) Two flows $\{\alpha_t\}_{t\in\mathbb{R}}$ and $\{\beta_t\}_{t\in\mathbb{R}}$ are said to be outer equivalent -- denoted by $\alpha \overset{\circ}{\sim} \beta$ -- if there exists a strongly continuous map $t \to u_t$ from $\mathbb{R}$ to $\mathcal{U}(M)$ such that, for all $s,t$ in $\mathbb{R}$ and $x$ in $M$, $u_{t+s} = u_t \alpha_t(u_s)$ and $\beta_t(x) = u_t \alpha_t(x) u_t^*$.

(c) A flow $\{\alpha_t\}$ is said to be inner if it is outer equivalent to the trivial flow (given by $\epsilon_t(x) = x \ \forall x \in M, t \in \mathbb{R}$), or, equivalently, if $\alpha_t(x) = u_t x u_t^*$ where $\{u_t\}$ is a strongly continuous one-parameter group of unitary operators in $M$. □

Hence the unitary cocycle theorem asserts that the modular groups corresponding to any pair of fns weights on $M$ are outer equivalent. Some simple exercises will help to clarify these concepts.

## Exercises

**(3.1.3)**

(a) Outer equivalence is an equivalence relation on the set of flows. (Hint: if $\{u_t\}$: $\{\alpha_t\} \overset{\circ}{\sim} \{\beta_t\}$, then $\{u_t^*\}$: $\{\beta_t\} \overset{\circ}{\sim} \{\alpha_t\}$ and, although taking adjoints is not strongly continuous, it is weakly continuous and hence, when restricted to $\mathcal{U}(M)$, it is strongly continuous.)

(b) If $\{\alpha_t\}$ is a flow on $M$ and if $t \to u_t$ is a strongly continuous map from $\mathbb{R}$ to $\mathcal{U}(M)$, let $\beta_t(x) = u_t \alpha_t(x) u_t^*$; show that if $\{u_t\}$ satisfies $u_{t+s} = u_t \alpha_t(u_s)$, then $\{\beta_t\}$ is a flow on $M$, which is necessarily outer equivalent to $\{\alpha_t\}$.

**(3.1.4)** Let $\phi$, $\psi$, $\{u_t\}$ be as in (the statement of) Theorem 3.1.1.

(a) If $\{w_t\}_{t \in \mathbb{R}}$ is a strongly continuous one-parameter group of unitary operators in $Z(M)$, and if $v_t = w_t u_t$, show that $\{v_t\}$ is a strongly continuous path of unitary operators in $M$, which also satisfies $v_{t+s} = v_t \sigma_t^\phi(v_s)$ and $\sigma_t^\psi(x) = v_t \sigma_t^\phi(x) v_t^*$, for $x \in M$, $s,t \in \mathbb{R}$

(Hint: you will need to use $Z(M) \subseteq M^\phi$.)

(b) If, conversely, $t \to v_t$ is a strongly continuous map from $\mathbb{R}$ to $\mathcal{U}(M)$ which also satisfies $v_{t+s} = v_t \sigma_t^\phi(v_s)$ and $\sigma_t^\psi(x) = v_t \sigma_t^\phi(x) v_t^*$, show tht there exists a strongly continuous one-parameter group $\{w_t\}$ in $\mathcal{U}(Z(M))$ such that $v_t = w_t u_t$ for all $t$. (Hint: put $w_t = u_t^* v_t$ and verify that $\{w_t\}$ is as wonderful as it is claimed to be.)

(c) If $M$ is a factor, the unitary cocycle of Theorem 3.1.1 is uniquely determined up to scaling by a continuous one-parameter group of complex scalars of unit modulus, i.e., if $\{u_t\}$ is one such unitary cocycle, any other unitary cocycle is of the form $v_t = e^{ita} u_t$ for some $a$ in $\mathbb{R}$.) □

In the remainder of this section, we shall discuss two results of Takesaki's: a part of the first result is a consequence of the cocycle theorem, while the second result explicitly produces a cocycle which works in some cases and also explains why the cocycle theorem is sometimes referred to as Connes' Radon-Nikodym Theorem. The proofs of both these results are somewhat technical, and we shall only present the proof under some additional hypothesis -- invariably that some self-adjoint operator is bounded. The theorems will be stated in their full generality, while the simplifying assumption will be spelt out at an appropriate juncture in the proof. Before proceeding to these results, however, we extend the notion of semifiniteness to a general, possibly non-factorial, von Neumann algebra.

**Definition 3.1.5.** A von Neumann algebra is said to be:

(a) semifinite, if it admits a fns trace;
(b) finite, if it admits a faithful normal tracial state. □

**Theorem 3.1.6.** *The following conditions on $M$ are equivalent:*

(i) *$M$ is semifinite;*
(ii) *the flow $\{\sigma_t^\phi\}$ is inner, for some* fns *weight $\phi$ on $M$;*

(iii) *the flow $\{\sigma_t^\phi\}$ is inner, for every* fns *weight $\phi$ on $M$.*

**Proof.** (i) ⇒ (ii): By assumption, there exists a fns trace $\tau$ on $M$; then $\{\sigma_t^\tau\}$ is the trivial flow on $M$ and hence (trivially) inner.

**(ii) ⇒ (iii).** Let $\phi$ and $\psi$ be fns weights on $M$, and suppose $\sigma^\phi$ is inner; thus $\sigma^\phi \overset{\circ}{\sim} \epsilon$, where $\epsilon$ denotes the trivial flow. By Theorem 3.1.1 ($\sigma^\psi \overset{\circ}{\sim} \sigma^\phi$) and Ex. (3.1.3) (a), conclude that $\sigma^\psi \overset{\circ}{\sim} \epsilon$ and hence $\sigma^\psi$ is inner.

**(iii) ⇒ (i).** Fix a fns weight $\phi$ on $M$ and assume, with no loss of generality, that $\mathcal{H} = \mathcal{H}_\phi$ and $\pi_\phi(x) = x$ for $x$ in $M$. Let us simply write $\Delta$ for $\Delta_\phi$. The assumption that $\sigma^\phi$ is inner means that there exists a strongly continuous one-parameter unitary group $\{u_t\}$ in $M$ such that $u_t x u_t^* = \Delta^{it} x \Delta^{-it}$ for all $x$ in $M$ and $t$ in $\mathbb{R}$. Set $x = u_s$ to infer that $u_s$ commutes with $\Delta^{it}$ for all $s$ and $t$; in other words $\{u_t\} \subseteq M^\phi$. Another consequence of this commutativity is that if $u_t' = u_t^* \Delta^{it}$, then $u_{t+s}' = u_t' u_s'$ for $s,t$ in $\mathbb{R}$. Thus $\{u_t'\}$ is a strongly continuous one parameter unitary group in $\mathcal{H}$; further $u_t' x u_t'^* = x$ for all $x$ in $M$, so that $u_t' \in M'$ for all $t$.

By Stone's theorem (and the double commutant theorem), there exist self-adjoint operators $H \,\eta\, M^\phi$ and $H' \,\eta\, M'$ such that $u_t = e^{itH}$ and $u_t' = e^{itH'}$ for all $t$ in $\mathbb{R}$. Since $\Delta^{it} = e^{itH}e^{itH'}$, and since $H$ and $H'$ commute, it follows that $\Delta^{1/2}$ is the closure of the operator $e^{H/2}e^{H'/2}$ -- for unbounded $A$ and $B$, $AB$ being defined naturally on dom $AB$ = dom $B \cap B^{-1}$ (dom $A$). (If $H$ and $H'$ are bounded, then $\Delta = e^{H+H'}$, and $\Delta^{1/2} = e^{H/2+H'/2} = e^{H/2}e^{H'/2}$; for unbounded $H$ and $H'$, the above equations are valid if the sum and product of two commuting self-adjoint operators are defined as the closures of the sum and product defined on the natural domains; a couple of exercises at the end of this theorem may clarify these matters to the reader who is not too comfortable with unbounded operators.)

Since $e^{-H}$ is an invertible positive self-adjoint operator affiliated to $M^\phi$, we may define $\tau = \phi(e^{-H}.)$ as in Theorem 2.6.3. The theorem will be proved once it has been established that this $\tau$ is a fns trace on $M$. We shall establish this under the assumption that both $H$ and $H'$ are bounded and $\phi$ is a finite weight.

So, suppose $\phi \in M_{*,+}$ and $H = h \in M^\phi$, $H' = h' \in M'$; since we are assuming that the action is taking place in the GNS space of $\phi$, there exists a cyclic and separating vector $\Omega$ for $M$ such that $\phi(x) = \langle x\Omega,\Omega\rangle$. Finally, we have defined $\tau(x) = \phi(e^{-h}x) = \phi(e^{-h/2}xe^{-h/2}) = \langle xe^{-h/2}\Omega, e^{-h/2}\Omega\rangle$ since $h \in M^\phi$. It follows from Ex. (2.6.1) that $\tau$ is a faithful normal positive linear functional. To verify that $\tau$ is tracial, pick $x,y$ in $M$ and compute:

$$\begin{aligned}
\tau(xy^*) &= \langle e^{-h/2}xy^*\, e^{-h/2}\Omega,\Omega\rangle \\
&= \langle y^*e^{-h/2}\Omega,\ x^*e^{-h/2}\Omega\rangle \\
&= \langle J\Delta^{1/2}(e^{-h/2}y\Omega),\ J\Delta^{1/2}(e^{-h/2}x\Omega)\rangle \\
&= \langle \Delta^{1/2}e^{-h/2}x\Omega,\ \Delta^{1/2}e^{-h/2}y\Omega\rangle \\
&= \langle e^{h'/2}x\Omega,\ e^{h'/2}y\Omega\rangle
\end{aligned}$$

$$= \langle e^{h'}x\Omega, y\Omega\rangle \; ; \qquad (*)$$

so,

$$\begin{aligned}\tau(xy) &= \tau((xy)1^*) \\ &= \langle e^{h'}xy\Omega,\Omega\rangle && \text{(by (*))} \\ &= \langle e^{h'}y\Omega, x^*\Omega\rangle && \text{(since } h' \in M'\text{)} \\ &= \tau(yx) && \text{(again by (*))},\end{aligned}$$

and the proof is complete. □

## Exercises

**(3.1.7)** Let $\{e_n\}_{n=1}^{\infty}$ and $\{f_n\}_{n=1}^{\infty}$ be sequences of projections such that $e_n \nearrow 1$ and $f_n \nearrow 1$.

(a) If $e_n$ and $f_n$ commute for all $n$, then $e_n \wedge f_n \nearrow 1$. (Hint: $e_n \wedge f_n = e_n f_n$ under the hypothesis, and multiplication is jointly continuous in the strong topology, on bounded sets.)
(b) Show that $e_n \wedge f_n$ may be 0 for all $n$, if the hypothesis of commutativity is dropped. (Hint: in $\mathcal{H} = L^2[0,1]$, let ran $e_n = L^2[1/n, 1]$ and ran $f_n$ = set of polynomials of degree $< n$; non-zero polynomials cannot vanish too often.)

**(3.1.8)** Let $\mathcal{M}_1 \subseteq \mathcal{M}_2 \subseteq \cdots$ be an increasing sequence of closed subspaces, whose union is dense in $\mathcal{H}$. If $T$ is a closed operator such that dom $T \supseteq \cup\mathcal{M}_n$, $T\mathcal{M}_n \subseteq \mathcal{M}_n$ and $T^*\mathcal{M}_n \subseteq \mathcal{M}_n$ for all $n$, show that:

(a) $T|\mathcal{M}_n$ is bounded for each $n$. (Hint: closed graph theorem.)
(b) $\cup\mathcal{M}_n$ is a core for $T$. (Hint: if $p_n = p_{\mathcal{M}_n}$, then $p_n \nearrow 1$ and $p_nT \subseteq Tp_n$.)

**(3.1.9)** Let $A$ and $B$ be self-adjoint operators in $\mathcal{H}$ such that $1_E(A)$ and $1_F(B)$ commute, for all Borel subsets $E$ and $F$ of $\mathbb{R}$ (for example, $A = H$, $B = H'$ as in the proof of Theorem 3.1.6). Let $e_n = 1_{[-n,n]}(A)$, $f_n = 1_{[-n,n]}(B)$, $p_n = e_n \wedge f_n$.

(a) $p_n \nearrow 1$, $p_nA \subseteq Ap_n$ and $p_nB \subseteq Bp_n$ for all $n$. (Hint: $e_nA$ is bounded and commutes with $f_n$.)
(b) $\mathcal{D}_0 = \cup$ ran $p_n$ is a core for $f(A)$ as well as for $g(B)$, for any two continuous functions $f$ and $g$ on $\mathbb{R}$; further, $f(A)\mathcal{D}_0 \cup g(B)\mathcal{D}_0 \subseteq \mathcal{D}_0$. (Hint: Apply Ex. (3.1.8) to each of $f(A)$ and $g(B)$, with $\mathcal{M}_n$ = ran $p_n$.)

(c) Define $(A + B)\xi = A\xi + B\xi$ if $\xi \in \text{dom}(A + B) = \text{dom}\, A \cap \text{dom}\, B$. Show that:

(i) $A + B$ is densely defined;
(ii) $(A + B) \subseteq (A + B)^*$ and so $(A + B)$ is closable;
(iii) $(\overline{A+B})$ is self-adjoint and has $\mathcal{D}_0$ as a core. (Hint: apply Ex. (3.1.8) to all the operators in sight, with $\mathcal{M}_n = \text{ran}\, p_n$.)

(d) Define the product $HK$ of two unbounded operators naturally, with $\text{dom}(HK) = \text{dom}\, K \cap K^{-1}(\text{dom}\, H)$. Let $f$ and $g$ be continuous (complex-valued) functions on $\mathbb{R}$. Show that:

(i) $f(A)g(B)$ is densely defined;
(ii) If $\bar{f}$ and $\bar{g}$ denote the complex-conjugates of $f$ and $g$, then

$(f(A)g(B))^* \supseteq \bar{g}(B)\bar{f}(A)$, and so $f(A)g(B)$ is closable;
(iii) $\mathcal{D}_0$ is a core for $f(A)g(B)$. (Hint: see hint for (c)(iii).)

(e) For any $z$ in $\mathbb{C}$, show that

$$(\overline{e^{zA}e^{zB}}) = e^{z(\overline{A+B})}.$$

(In the context of Theorem 3.1.6, this says that

$$\Delta^{it} = e^{itH}\, e^{itH'} = e^{it(\overline{H+H'})},$$

so that

$$\Delta = e^{(\overline{H+H'})}$$

and

$$\Delta^{1/2} = e^{1/2(\overline{H+H'})} = (\overline{e^{1/2H}e^{1/2H'}}).) \qquad \square$$

**Theorem 3.1.10.** *Let $\phi$ be a* fns *weight on $M$. Let $\psi$ be another* fns *weight on $M$ such that $\psi \circ \sigma_t^\phi = \psi$ for all $t$. Then $\sigma_t^\psi(x) = H^{it}\sigma_t^\phi(x)H^{-it}$ for all $t$ in $\mathbb{R}$ and $x$ in $M$, where $H$ is (the Radon-Nikodym density $d\psi/d\phi$) as in Theorem* 2.6.3.

**Proof.** Assume, without loss of generality that $\mathcal{H} = \mathcal{H}_\phi$ and $\pi_\phi(x) = x$ for all $x$ in $M$. We shall simply write $J$, $S$ and $\Delta$ for $J_\phi$, $S_\phi$ and $\Delta_\phi$. Since $H \,\eta\, M^\phi$, it follows that $\Delta^{it} H \Delta^{-it} = H$ for all $t$. It follows (since $H$ is an invertible positive self-adjoint operator) that $\Delta^{it}$ and $H^{is}$ commute for all $s$ and $t$. Consequently $\{H^{it}\Delta^{it}\}_{t\in\mathbb{R}}$ is a strongly continuous one-parameter unitary group in $\mathcal{H}$. Since $H^{it} \in M$, conclude that the equation $\alpha_t(x) = H^{it}\Delta^{it}x\Delta^{-it}H^{-it}$ defines a flow on $M$; since $\alpha_t(x) = H^{it}\sigma_t^\phi(x)H^{-it}$, the theorem will be proved the moment we establish that $\psi$ satisfies the KMS condition with respect to $\{\alpha_t\}$. We shall establish this under the further assumptions that (i) $\phi,\psi \in M_{*,+}$ and (ii) the operators $\Delta$, $H$ and $H^{-1}$ are bounded. (Note that the boundedness of $H$ and $H^{-1}$ amounts to the inequality $c_1\phi \leqslant \psi \leqslant c_2\phi$ for some constants $c_1, c_2 > 0$.)

Consistent with our notational convention, let us write $H = h \in M^{\phi}$. The assumption that all the action is going on in $\mathcal{H} = \mathcal{H}_{\phi}$ means that there is a cyclic and separating vector $\Omega$ for $M$ such that $\phi(x) = \langle x\Omega,\Omega\rangle$, and consequently $\psi(x) = \langle hx\Omega,\Omega\rangle$. Now let $x,y \in M$ and define $F\colon \mathbb{C} \to \mathbb{C}$ by $F(z) = \langle h^{iz+1}\Delta^{iz+1}y\Omega, \Delta^{-1/2}Jh^{-iz}x\Omega\rangle$. Since, by hypothesis, the spectra of $h$ and $\Delta$ are a safe distance away from the origin, it follows that $F$ is an entire function. (Note that the $J$ factor in the second term "cancels" the conjugate linearity of the inner product in the second variable.) Some easy estimates, of the sort used in the proof of the Tomita-Takesaki theorem for finite weights (Th. 2.3.3), show that $F$ is bounded in every strip of the form $|\mathrm{Im}\, z| \leqslant \gamma$. Finally, compute:

$$\begin{aligned} F(t) &= \langle h^{it+1}\Delta^{it+1}y\Omega, \Delta^{-1/2}Jh^{-it}x\Omega\rangle \\ &= \langle \Delta h^{it+1}\Delta^{it}y\Omega, \Delta^{-1/2}Jh^{-it}x\Omega\rangle \\ &= \langle h^{-it}x\Omega, (J\Delta^{1/2})(h^{it+1}\sigma_t^{\phi}(y)\Omega)\rangle \\ &= \langle h^{-it}x\Omega, \sigma_t^{\phi}(y^*)h^{-it+1}\Omega\rangle \\ &= \phi(h.h^{it}\,\sigma_t^{\phi}(y)h^{-it}.x) \\ &= \psi(\alpha_t(y)x), \end{aligned}$$

while

$$\begin{aligned} F(t+i) &= \langle h^{it}\Delta^{it}y\Omega, (\Delta^{-1/2}J)h^{-it+1}x\Omega\rangle \\ &= \langle h^{it}\Delta^{it}y\Omega, x^*h^{it+1}\Omega\rangle && (\text{since } \Delta^{-1/2}J = S) \\ &= \phi(h^{-it}h\; xh^{it}\;\sigma_t^{\phi}(y)) \\ &= \phi(h.xh^{it}\sigma_t^{\phi}(y)h^{-it}) && (\text{since } h^{-it} \in M^{\phi}) \\ &= \psi(x\alpha_t(y)); \end{aligned}$$

in other words, $F$ is KMS-admissible for $x$ and $y$ (relative to $\psi$) and the theorem is proved, at least in the special case we have considered. □

In view of Ex. (3.1.4) (c), if $\phi$ and $\psi$ are fns weights on a factor, and if $\psi$ is invariant under $\sigma^{\phi}$, then any unitary cocycle $\{u_t\}$ as in Theorem 3.1.1 must be of the form $u_t = e^{it\lambda} H^{it}$ for some $\lambda$ in $\mathbb{R}$, where $H$ is the "Radon-Nikodym derivative of $\psi$ with respect to $\phi$" in the sense of Theorem 2.6.3. It follows that if one has managed to find one unitary cocycle $\{u_t\}$, then $\{u_t\}$ is actually a one-parameter group and the "density" $H$ is determined, up to multiplication by a positive scalar, as the exponential of the infinitesimal generator of the group $\{u_t\}$. Hence there is a strong case for calling Theorem 3.1.1 a Radon-Nikodym theorem.

## 3.2. The Arveson Spectrum of an Action

Throughout this section and the next, the symbol $G$ will denote a (Hausdorff) locally compact abelian group, with dual (or character) group $\Gamma$. While it is true that, as far as our application is concerned, only the case $G = \mathbb{R}$ is relevant, we shall persist with the abstract situation for the following reasons: (a) when $G = \mathbb{R}$, the dual $\Gamma$ also becomes identical with $\mathbb{R}$, the consequent identification of $G$ with $\Gamma$ being not particularly desirable; and (b) the proofs are no harder for a general $G$ than for $\mathbb{R}$, and no student has been irreparably harmed for having had to learn some abstract harmonic analysis. (For the reader who is unfamiliar with a non-empty subset of {locally compact group, character group, Haar measure, Fourier transform}, there is a brief appendix devoted to these notions.)

We shall assume that $G$ satisfies the second axiom of countability. One consequence is that $G$ is metrizable and separable, so that, in considering $L^2(G)$, we shall still stay in the category of separable Hilbert spaces; another consequence is that all open sets in $G$ are $\sigma$-compact -- i.e., countable unions of compact sets -- so that there need be no fuss about Baire sets and Borel sets. The class of Borel sets is the smallest $\sigma$-algebra $\mathcal{F}_G$ containing all the compact sets in $G$; this $\sigma$-algebra will be the domain of definition of all measures we shall consider. The space of finite, regular, complex measures on $G$ is denoted by $M(G)$; the set $M(G)$ has the structure of an involutive Banach algebra, when equipped with the involution $\mu^*(E) = \overline{\mu(-E)}$, total-variation norm

$$\left[\|\mu\| = \sup\left\{\sum_{i=1}^{n} |\mu(E_i)|: \{E_i\}_{i=1}^{n} \text{ a Borel partition of } G\right\}\right]$$

and convolution product $(\mu * \nu)(E) = \int \mu(A - t)d\nu(t)$.

Once and for all, fix Haar measures on $G$ and $\Gamma$ -- denoted simply by $dt$ and $d\gamma$ -- the normalizations so chosen that the Fourier-Plancherel transform $f \to \hat{f}$ is a unitary operator from $L^2(G)$ to $L^2(\Gamma)$. The space $L^1(G)$ can be identified with the closed ideal in $M(G)$ consisting of measures absolutely continuous with respect to Haar measure. In particular, $L^1(G)$ is an involutive commutative Banach algebra with respect to $\|\cdot\|_1$ and the operations $f * g(t) = \int f(s)g(t-s)ds$ and $f^*(t) = \overline{f(-t)}$.

Stone's theorem states that if $t \to u_t$ is a strongly continuous unitary representation of $G$ in a Hilbert space $\mathcal{H}$, then there exists a spectral measure $e: \mathcal{F}_\Gamma \to \mathcal{P}(\mathcal{L}(\mathcal{H}))$ such that $u_t = \int \langle t,\gamma\rangle^{-1} de(\gamma)$ (the duality between $G$ and $\Gamma$ expressed by $(t,\gamma) \to \langle t,\gamma\rangle$). For $f$ in $L^1(G)$, if we define $u(f) = \int \hat{f}(\gamma)de(\gamma)$, it follows that $f \to u(f)$ is a *-algebra homomorphism of $L^1(G)$ into $\mathfrak{L}(\mathcal{H})$; it is not hard to see that $u(f) = \int f(t)u_t dt$, in the strong sense of the integral. It can, further, be seen that ran $e(E) = \{\xi \in \mathcal{H}: u(f)\xi = 0$ whenever $f \in L^1(G)$ and $\hat{f}$ vanishes on $E\}$, for any closed set $E$ in $\Gamma$. Our aim, in this section, is to carry out a similar analysis of a group action on a von Neumann algebra.

**Definition 3.2.1.** (a) An action $\alpha$ of $G$ on a von Neumann algebra $M$ is a homomorphism $\alpha$ from $G$ into the group Aut $M$ of *-automorphisms of $M$, which is pointwise $\sigma$-weakly continuous -- i.e., $t \to \alpha_t(x)$ is $\sigma$-weakly continuous, for each $x$ in $M$.

(b) A dynamical system is a triple $(M,G,\alpha)$ consisting of a von Neumann algebra $M$, a locally compact group $G$, and an action $\alpha$ of $G$ on $M$. □

Notice that we have written $\alpha_t$ rather than $\alpha(t)$ for the automorphism associated with it; this will cut down the number of parentheses to be used. For the same reason, we shall write $\langle x,\phi\rangle$ for $\phi(x)$ and $\langle t,\gamma\rangle$ for $\gamma(t)$.

**Proposition 3.2.2.** *Let $\alpha$ be an action of $G$ on $M$. The equation $\alpha(\mu) = \int \alpha_t d\mu(t)$ -- the integral being interpreted in the pointwise $\sigma$-weak sense -- induces an algebra homomorphism $\mu \to \alpha(\mu)$ from $M(G)$ into the algebra $\mathfrak{L}_\sigma(M)$ of $\sigma$-weakly continuous linear operators on $M$. Further, $\|\alpha(\mu)\| \leq \|\mu\|$.*

**Proof.** Fix $\mu$ in $M(G)$, $x$ in $M$, $\phi$ in $M_*$. Recall that every *-automorphism of $M$ is isometric (and $\sigma$-weakly continuous); so $t \to \langle\alpha_t(x),\phi\rangle$ is a bounded (by $\|x\|\ \|\phi\|$) continuous function on $G$ (by the definition of an action). It follows that the assignment

$$\phi \to \int \langle\alpha_t(x),\phi\rangle d\mu(t)$$

defines a bounded linear functional on $M_*$ of norm at most $\|x\|\ \|\mu\|$; since $(M_*)^* = M$, there exists a unique element in $M$, which we shall denote by $\alpha(\mu)x$, such that $\|\alpha(\mu)x\| \leq \|\mu\|\ \|x\|$ and

$$\langle\alpha(\mu)x,\phi\rangle = \int\langle\alpha_t(x),\phi\rangle\ d\mu(t)$$

for all $\phi$ in $M_*$. Clearly, the assignment $x \to \alpha(\mu)x$ is linear and so $\alpha(\mu) \in \mathfrak{L}(M)$ and $\|\alpha(\mu)\| \leq \|\mu\|$. It is equally clear that the map $\mu \to \alpha(\mu)$ is linear.

To prove $\alpha(\nu * \mu) = \alpha(\nu)\alpha(\mu)$ proceed thus: if $s \in G$, $x \in M$ and $\phi \in M_*$, note first that $\phi \circ \alpha_s \in M_*$ (since $\alpha_s$ is $\sigma$-weakly continuous) and that $\langle\alpha_s(\alpha(\mu)x),\phi\rangle = \langle\alpha(\mu)x,\ \phi \circ \alpha_s\rangle = \int\langle\alpha_t(x),\phi \circ \alpha_s\rangle d\mu(t) = \langle\alpha(\mu)(\alpha_s(x)),\phi\rangle$ and hence $\alpha_s \circ \alpha(\mu) = \alpha(\mu) \circ \alpha_s$. Note next that, by definition of convolution,

$$\int g(t)d(\nu * \mu)(t) = \iint g(t + s)d\nu(s)d\mu(t)$$

for any bounded measurable function $g$ on $G$; hence,

$$\begin{aligned} \langle\alpha(\nu * \mu)x,\phi\rangle &= \iint \langle\alpha_{s+t}(x),\phi\rangle d\nu(s)d\mu(t) \\ &= \int \langle\alpha(\mu)(\alpha_s(x)),\phi\rangle\, d\nu(s) \\ &= \int \langle\alpha_s(\alpha(\mu)x),\phi\rangle\, d\nu(s) \\ &= \langle\alpha(\nu)\alpha(\mu)x,\phi\rangle . \end{aligned}$$

To complete the proof, we need to verify that each $\alpha(\mu)$ is $\sigma$-weakly continuous; it clearly suffices to do this for a positive measure $\mu$. We need to show that $\phi \in M_*$ implies $\phi \circ \alpha(\mu) \in M_*$; for this, too, we may assume that $\phi$ is positive. It follows from $\mu \geqslant 0$, $\phi \geqslant 0$ that $\phi \circ \alpha(\mu) \geqslant 0$, since for $x$ in $M_+$,

$$\langle x, \phi \circ \alpha(\mu)\rangle = \int \langle\alpha_t(x),\phi\rangle\, d\mu(t) \geqslant 0.$$

Suppose now that $\{x_i\}$ is a monotone net in $M_+$ and that $x_i \nearrow x$. Then, the net $\{\langle\alpha_.(x_i),\phi\rangle\}$ of continuous functions increases pointwise to the continuous function $\langle\alpha_.(x),\phi\rangle$. By Dini's theorem, the convergence must be uniform on compact subsets of $G$; it follows that for compact $K \subseteq G$,

$$\int_K \langle\alpha_t(x_i),\phi\rangle d\mu(t) \nearrow \int_K \langle\alpha_t(x),\phi\rangle d\mu(t).$$

As $G$ is $\sigma$-compact, it is easily deduced that the above statement remains valid with $K$ replaced by $G$. Hence $\langle\alpha(\mu)x_i,\phi\rangle \nearrow \langle\alpha(\mu)x,\phi\rangle$. Thus, $\phi \circ \alpha(\mu)$ is a normal positive linear functional and consequently $\sigma$-weakly continuous (cf. the comments following Ex. (0.4.5)). Finally, the proof is complete. □

If $f \in L^1(G)$ and $d\mu_f = f\,dt$, we shall simply write $\alpha(f)$ for $\alpha(\mu_f)$; thus

$$\langle\alpha(f)x,\phi\rangle = \int f(t) \langle \alpha_t(x),\phi\rangle dt.$$

## Exercises

(3.2.3) Let $\alpha$ be an action of $G$ on $M$.

(a) If $x \in M$ and $\alpha(f)x = 0$ for all $f$ in $C_c(G)$ (the space of continuous functions with compact support), then $x = 0$. (Hint: $C_c(G)$ is dense in $L^1(G)$ and the dual of $L^1(G)$ is $L^\infty(G)$.)

(b) $\{\phi \circ \alpha(f)\colon \phi \in M_*,\ f \in C_c(G)\}$ is a norm - total subset of $M_*$. (Hint: if $M_{*,0}$ is the closed subspace generated by this set and if $M_{*,0} \subsetneq M_*$, appeal to the Hahn-Banach theorem and (a).)

(c) Let $\{k_i: i \in I\}$ be a bounded approximate identity for $L^1(G)$ -- i.e., $\{k_i\}$ is a net such that

$$\sup_i \|k_i\| < \infty \quad \text{and} \quad \lim_i \|k_i * f - f\| = 0$$

for all $f$ in $L^1(G)$. Then

$$\lim_i \|\phi \circ \alpha(k_i) - \phi\| = 0$$

for all $\phi$ in $M_*$. (Hint: as $\sup \|\alpha(k_i)\| < \infty$, it is enough to prove this for any total set of $\phi$'s; if $\phi = \psi \circ \alpha(f)$,

$$\|\phi \circ \alpha(k_i) - \phi\| \leq \|\psi\| \, \|\alpha(f)\alpha(k_i) - \alpha(f)\| \leq \|\psi\| \, \|f * k_i - f\|;$$

also $L^1(G)$ is commutative.)

(d) If $\{k_i\}_{i \in I}$ is a bounded approximate identity for $L^1(G)$, then $\alpha(k_i)x \to x$ $\sigma$-weakly for each $x$ in $M$. □

We shall use the symbol $\hat{C}$ to denote the set $\hat{C} = \{f \in L^1(G): \hat{f}$ has compact support$\}$. It is hoped that the notation $\hat{C}$ will suggest the phrase "compactly supported Fourier transform" to the reader. Since $(f * g)\hat{} = \hat{f}\,\hat{g}$, it is clear that $\hat{C}$ is an ideal in $L^1(G)$. It is well-known that $\hat{C}$ has a rich supply of functions. We gather together some results (which may be found in [Rud], for instance) which we shall need. (Parts (a), (b) and (c) of the following proposition are, respectively, Theorems 2.6.2, 2.6.8 and 2.6.6 in [Rud].)

**Proposition 3.2.4.**

(a) *If $K \subseteq U \subseteq \Gamma$, with $K$ compact and $U$ open, there exists $f$ in $\hat{C}$ such that $1_K \leq \hat{f} \leq 1_U$.*

(b) *If $K$ is a compact set in $\Gamma$, and if $\epsilon > 0$, there exists $k$ in $\hat{C}$ such that $\|k\|_1 < 1 + \epsilon$ and $\hat{k} \equiv 1$ on $K$.*

(c) *If $f \in L^1(G)$, and $\epsilon > 0$, there exists $k$ in $\hat{C}$ such that $\|f - k * f\|_1 < \epsilon$; in particular, $\hat{C}$ is norm-dense in $L^1(G)$.* □

## Exercises

(3.2.5) Let $\Lambda = \{k \in \hat{C}: \|k\|_1 < 2\}$.

(a) If $k_1, k_2 \in \Lambda$, say that $k_1 \prec k_2$ if $\hat{k}_2 \equiv 1$ on the support of $\hat{k}_1$. Show that $\Lambda$ is directed upwards by this order relation; i.e., if $k_1, k_2 \in \Lambda$, there exists $k_3 \in \Lambda$ such that $k_i \prec k_3$ for $i = 1,2$. (Hint: apply Prop. 3.2.4 (b) with $\epsilon = 1$ and $K$ any compact set containing the supports of both $\hat{k}_1$ and $\hat{k}_2$.)

(b) Show that $\Lambda$ is a bounded approximate identity for $L^1(G)$. (Hint: since $\|k\| < 2$ for all $k$ in $\Lambda$, it is enough to prove that

$$\lim_{k \in \Lambda} \|k_* f - f\| = 0$$

for a dense set of $f$'s; if $f \in \hat{C}$, note that there exists $k_0$ in $\Lambda$ such that $k * f = f$ for all $k$ in $\Lambda$ such that $k_0 \leq k$; for this, you need to use the injectivity of the Fourier transform.) □

There is a natural way to pass from closed ideals of $L^1(G)$ to closed subsets of $\Gamma$ and vice versa. (In fact, this idea underlies the so-called hull-kernel topology.) To be precise, for a set $S \subseteq L^1(G)$, define the annihilator $S^\perp = \{\gamma \in \Gamma: \hat{f}(\gamma) = 0$ for all $f$ in $S\}$. Since $\hat{f}$ is continuous for any $f$ in $L^1(G)$, the set $S^\perp$ is always a closed set in $\Gamma$.

Conversely, for any subset $E$ of $\Gamma$, let $I(E) = \{f \in L^1(G): \hat{f}(E) = \{0\}\}$. It is clear that $I(E) = I(\bar{E})$ and that $I(E)$ is a closed ideal in $L^1(G)$. If $\gamma \notin \bar{E}$, Prop. 3.2.4 (a) shows that there exists $f$ in $L^1(G)$ such that $\hat{f}(\gamma) = 1$ and $\hat{f}(E) = 0$; consequently $\gamma \notin I(E)^\perp$. Since $\bar{E} \subseteq I(E)^\perp$ (clearly!), we have shown that $\bar{E} = I(E)^\perp$.

In the dual situation, when $S \subseteq L^1(G)$, it is clear that $S^\perp = I_S^\perp$, where $I_S$ is the smallest closed ideal containing $S$, and that $I_S \subseteq I(S^\perp)$. This inclusion can, however, be strict. This phenomenon is related to the so-called problem of spectral synthesis. A closed set $E$ in $\Gamma$ is said to have spectral synthesis if $I(E)$ is the only closed ideal in $L^1(G)$ with annihilator $E$. The celebrated Wiener Tauberian theorem states that singleton sets have spectral synthesis. In practice, the lack of spectral synthesis is often compensated for by the following result (cf. [Loo]).

**Proposition 3.2.6.** (a) *If $E$ is a closed set in $\Gamma$, then $I_0(E) = \{f \in L^1(G)$: $\hat{f}$ vanishes on a neighborhood of $E\}$ is an ideal in $L^1(G)$ and $E = I_0(E)^\perp$.*

(b) *If $I$ is a closed ideal in $L^1(G)$ and if $f \in L^1(G)$ is such that $\hat{f}$ vanishes on a neighborhood of $I^\perp$, then $f \in I$.*

(a) is proved as was $\bar{E} = I(E)^\perp$ above; (b) is a theorem in Section 37.C of [Loo]. □

**Definition 3.2.7.** Let $\alpha$ be an action of $G$ on $M$.

(a) Define the Arveson spectrum of $\alpha$ by

$$\text{sp}\,\alpha = \{f \in L^1(G): \alpha(f) = 0\}^\perp = \{\gamma \in \Gamma: \alpha(f) = 0 \Rightarrow \hat{f}(\gamma) = 0\}.$$

(b) If $x \in M$, let

$$\text{sp}_\alpha(x) = \{f \in L^1(G): \alpha(f)x = 0\}^\perp = \{\gamma \in \Gamma: \alpha(f)x = 0 \Rightarrow \hat{f}(\gamma) = 0\}.$$

(c) If $E$ is a closed subset of $\Gamma$, define the associated "spectral subspace" by

$$M(\alpha,E) = \{x \in M: \text{sp}_\alpha(x) \subseteq E\}. \quad □$$

**Lemma 3.2.8.** *Let $x \in M$ and let $E$ be a closed set in $\Gamma$.*

(a) $x \in M(\alpha,E) \Leftrightarrow \alpha(f)x = 0$ *whenever* $f \in L^1(G)$ *is such that* $\hat{f}$ *vanishes on a neighborhood of* $E$ (*i.e.,* $E \subseteq \text{Int } \{f\}^{\perp}$, *where* Int *denotes interior*); *consequently* $M(\alpha,E)$ *is a* $\sigma$*-weakly closed subspace of* $M$;

(b) $\text{sp}_{\alpha}(x) = \phi \Leftrightarrow x = 0$;

(c) $\text{sp}_{\alpha}(x) = \{0\} \Leftrightarrow \alpha_t(x) = x$ *for all* $t$ *in* $G$ (*i.e.,* $x \in M^{\alpha}$), *and* $x \neq 0$.

**Proof.** Let $I_x = \{f \in L^1(G): \alpha(f)x = 0\}$. Clearly $I_x$ is a closed ideal in $L^1(G)$ -- recall that $L^1(G)$ inherits commutativity from $G$. Further, by definition, $\text{sp}_{\alpha}(x) = I_x^{\perp}$.

(a) The implications ($\Rightarrow$) and ($\Leftarrow$) are easy consequences of (b) and (a) of Prop. 3.2.6 applied to the closed ideal $I_x$ and the closed set $E$, respectively. The second assertion follows from the first in view of the $\sigma$-weak continuity of the $\alpha(f)$'s.

(b) If $\text{sp}_{\alpha}(x) = \phi$, then by (a) (applied to $E = \phi$), we find that $\alpha(f)x = 0$ for all $f$ in $L^1(G)$, and so, by Ex. (3.2.3) (a), $x = 0$; it is clear, using Prop. 3.2.4, that $\text{sp}_{\alpha}(0) = \phi$.

(c) If $I_x^{\perp}$ ($= \text{sp}_{\alpha}(x)$) $= \{0\}$, it follows from Wiener's Tauberian theorem that $I_x = \hat{I}(\{0\}) = \{f \in L^1(G): \hat{f}(0) = 0\}$. Let $\Lambda$ be the bounded approximate identity for $L^1(G)$ constructed as in Ex. 3.2.5. It is clear that the set $\Lambda_0 = \{k \in \Lambda: \hat{k}(0) = 1\}$ is co-final in $\Lambda$ -- in fact, $k_1,k_2 \in \Lambda$, $k_1 < k_2$ and $k_1 \in \Lambda_0$ imply $k_2 \in \Lambda_0$. Thus $\Lambda_0$ is also a bounded approximate identity for $L^1(G)$ and so, by Ex. (3.2.3) (d),

$$x = \sigma\text{-weak} - \lim_{k \in \Lambda_0} \alpha(k)x.$$

If, now, $f \in L^1(G)$, note that $(f - \hat{f}(0)k)\hat{}(0) = 0$ for all $k$ in $\Lambda_0$ and so $\alpha(f)x = \hat{f}(0)x$. Hence, for all $f$ in $L^1(G)$ and $\phi$ in $M_*$,

$$\int f(t)\langle\alpha_t(x) - x,\phi\rangle\, dt = \langle\alpha(f)x - \hat{f}(0)x,\phi\rangle = 0.$$

Since $f$ is arbitrary, conclude -- since the dual of $L^1(G)$ is $L^{\infty}(G)$ -- that $\langle\alpha_t(x) - x,\phi\rangle = 0$ almost everywhere (with respect to Haar measure). Since the function $t \to \langle\alpha_t(x) - x,\phi\rangle$ is continuous, conclude that $\langle\alpha_t(x) - x,\phi\rangle = 0$ for every $t$. Since $\phi$ is arbitrary, conclude that $x \in M^{\alpha}$. Also, by (b), $x \neq 0$.

Conversely, if $\alpha_t(x) = x$ for all $t$, it follows that

$$\alpha(f)x \;(= \int f(t)\alpha_t(x)dt) = \hat{f}(0)x$$

for any $f$ in $L^1(G)$. So, if $x \neq 0$, $I_x = I(\{0\})$ and so $\text{sp}_{\alpha}(x) = I(\{0\})^{\perp} = \{0\}$. □

The following proposition lists some further elementary facts concerning the notions introduced in Definition 3.2.7; we use the notation spt $\hat{f}$ for the support of the function $\hat{f}$: spt $\hat{f} = \overline{\hat{f}^{-1}(\mathbb{C} \setminus \{0\})}$.

**Proposition 3.2.9.** *Let $f \in L^1(G)$, $x \in M$, and $E$ be a closed subset of $\Gamma$.*

(a) $\mathrm{sp}_\alpha(x^*) = -\,\mathrm{sp}_\alpha(x)$;
(b) *If $\mathcal{B}$ is any $\sigma$-weakly total subset of $M$, then*

$$\mathrm{sp}\,\alpha = \Big[\bigcup_{y\in\mathcal{B}} \mathrm{sp}_\alpha(y)\Big]^- ;$$

(c) $\alpha_t(M(\alpha,E)) = M(\alpha,E)$ *for all $t$ in $G$;*
(d) $\mathrm{sp}_\alpha(\alpha(f)x) \subseteq \mathrm{sp}_\alpha(x) \cap \mathrm{spt}\,\hat{f}$;
(e) $\gamma \in \mathrm{sp}\,\alpha \Leftrightarrow M(\alpha,V) \neq \{0\}$ *for every neighborhood $V$ of $\gamma$.*
(f) *If $\mu \in M(G)$ and $\hat{\mu}$ vanishes identically on a neighborhood of $\mathrm{sp}_\alpha(x)$, then $\alpha(\mu)x = 0$.*

**Proof.** (a) This follows from the easily verified equations

$$(\alpha(f)x)^* = \alpha(\bar{f})x^* \quad \text{and} \quad \hat{\bar{f}}(\gamma) = \overline{\hat{f}(-\gamma)} .$$

(b) Clearly $\alpha(f) = 0$ implies $\alpha(f)y = 0$ for all $y$ in $M$; it follows that $\mathrm{sp}\,\alpha \supseteq \mathrm{sp}_\alpha(y)$ for all $y$, and since $\mathrm{sp}\,\alpha$ is closed, we get

$$\mathrm{sp}\,\alpha \supseteq \Big[\bigcup_{y\in\mathcal{B}} \mathrm{sp}_\alpha(y)\Big]^- .$$

Conversely, if

$$\gamma \notin \Big[\bigcup_{y\in\mathcal{B}} \mathrm{sp}_\alpha(y)\Big]^- ,$$

pick $f$ in $L^1(G)$ such that $\hat{f}(\gamma) = 1$ and $\hat{f}$ vanishes on an open neighborhood of $(\bigcup_{y\in\mathcal{B}} \mathrm{sp}_\alpha(y))^-$. It follows from Lemma 3.2.8 (a) that $\alpha(f)y = 0$ for all $y$ in $\mathcal{B}$. The assumption on $\mathcal{B}$ and the $\sigma$-weak continuity of $\alpha(f)$ now ensure that $\alpha(f) = 0$. Since $\hat{f}(\gamma) \neq 0$, infer that $\gamma \notin \mathrm{sp}\,\alpha$.

(c) Note that $\alpha(f)\alpha_t(x) = \alpha(f_t)x$, where $f_t(s) = f(s - t)$; also $\hat{f_t}(\gamma) = \langle t,\gamma\rangle^{-1}\hat{f}(\gamma)$, so that $\hat{f}(\gamma) = 0 \Leftrightarrow \hat{f_t}(\gamma) = 0$. Consequently, it follows from Lemma 3.2.8 (a) that $x \in M(\alpha,E) \Leftrightarrow \alpha_t(x) \in M(\alpha,E)$.

(d) $\alpha(g)x = 0 \Rightarrow \alpha(g)(\alpha(f)x) = \alpha(g * f)(x) = \alpha(f * g)(x) = \alpha(f)\alpha(g)x = 0$; so $\mathrm{sp}_\alpha(x) \supseteq \mathrm{sp}_\alpha(\alpha(f)x)$. On the other hand, if $\gamma \notin \mathrm{spt}\,\hat{f}$, there exists (by Prop. 3.2.4 (a)) $g$ in $\hat{C}$ such that $\hat{g}(\gamma) = 1$ and $\hat{g}\hat{f} \equiv 0$. Conclude that $g * f = 0$ so that $\alpha(g)(\alpha(f)x) = 0$, while $\hat{g}(\gamma) \neq 0$; hence $\gamma \notin \mathrm{sp}_\alpha(\alpha(f)x)$.

(e) If $M(\alpha,V) = \{0\}$ for some neighborhood $V$ of $\gamma$, pick $g$ in $\hat{C}$ such that $\hat{g}(\gamma) = 1$ and $\mathrm{spt}\,\hat{g} \subseteq V$. Then, for any $x$ in $M$, part (d) implies that $\mathrm{sp}_\alpha(\alpha(g)x) \subseteq V$, whence $\alpha(g)x \in M(\alpha,V)$, and so $\alpha(g)x = 0$. Thus $\alpha(g) = 0$ while $\hat{g}(\gamma) \neq 0$; so $\gamma \notin \mathrm{sp}\,\alpha$.

Conversely, suppose $M(\alpha,V) \neq \{0\}$ for every neighborhood $V$ of $\gamma$. If $0 \neq x_V \in M(\alpha,V)$, then, by Lemma 3.2.8 (b), $\mathrm{sp}_\alpha(x_V) \neq \phi$; pick $\gamma_V \in \mathrm{sp}_\alpha(x_V) \subseteq V$. Clearly $\gamma_V \in \mathrm{sp}\,\alpha$, since $\mathrm{sp}_\alpha(x) \subseteq \mathrm{sp}\,\alpha$ for all $x$ in $M$.

So, $V \cap \mathrm{sp}\, \alpha \neq \phi$, for every open neighborhood $V$ of $\gamma$; since sp $\alpha$ is closed, conclude that $\gamma \in \mathrm{sp}\, \alpha$.

(f) If $f \in L^1(G)$, then $\widehat{f * \mu} = \hat{f}\, \hat{\mu}$ vanishes on a neighborhood of $\mathrm{sp}_\alpha(x)$ and so, by Lemma 3.2.8 (a), $\alpha(f)\alpha(\mu)x = 0$. Since $f$ was arbitrary, conclude from Ex. (3.2.3) (a) that $\alpha(\mu)x = 0$. □

We conclude this section with an analogue of the statement that $\mathrm{spt}(f * g) \subseteq \overline{\mathrm{spt}\ f + \mathrm{spt}\ g}$.

**Proposition 3.2.10.** *Let $E_1$ and $E_2$ be closed subsets of $\Gamma$ and let $E = \overline{E_1 + E_2}$. If $x_i \in M(\alpha,E_i)$ for $i = 1,2$, and $x = x_1x_2$, then $x \in M(\alpha,E)$.*

**Proof. Case (i).** $\mathrm{sp}_\alpha(x_i)$ is compact, for $i = 1,2$.

In view of Lemma 3.2.8 (a), we need to show that $\alpha(f)x = 0$ whenever $f \in L^1(G)$ is such that $\hat{f}$ vanishes in a neighborhood of $E$. Also, in the case under discussion, we may assume $E_1$ and $E_2$ are compact (by replacing $E_i$ by $\mathrm{sp}_\alpha(x_i)$); then $E = E_1 + E_2$ is also compact.

Let $V$ be a neighborhood of 0 in $\Gamma$ such that $\hat{f}$ vanishes on $E + V + V$. Appeal to Prop. 3.2.4 (a), and pick $f_i$ in $\hat{C}$ such that $\hat{f}_i$ is identically equal to one on a neighborhood of $E_i$ and $\mathrm{spt}\ \hat{f}_i \subseteq E_i + V$, for $i = 1,2$. (Locally compact Hausdorff spaces are regular!) Notice that, by Prop. 3.2.9 (f), $\alpha(f_i)x_i = x_i$, $i = 1,2$. So, for any $\phi$ in $M_*$,

$$\begin{aligned}
\langle\alpha(f)x,\phi\rangle &= \langle\alpha(f)\{(\alpha(f_1)x_1)(\alpha(f_2)x_2)\},\phi\rangle \\
&= \iiint f(t)f_1(t_1)f_2(t_2)\langle(\alpha_{t+t_1}(x_1))(\alpha_{t+t_2}(x_2)),\phi\rangle dt\, dt_1 dt_2 \\
&= \iiint f(s)f_1(s_1 - s)f_2(s_2 + s_1 - s)\cdot \\
&\qquad \cdot\langle(\alpha_{s_1}(x_1))(\alpha_{s_1+s_2}(x_2)),\phi\rangle ds\, ds_1 ds_2,
\end{aligned}$$

where we have used (i) Fubini's theorem and the $\sigma$-weak continuity of $\alpha(g)$ which ensures that $\alpha(g)$ may be "taken inside" a $\sigma$-weakly defined integral, in the second equality, and (ii) Fubini's theorem and the substitutions $s = t$, $s_1 = t + t_1$, $s_2 = t_2 - t_1$, in the third equality. Hence

$$\langle\alpha(f)x,\phi\rangle = \iint k(s_1,s_2)h(s_1,s_2)ds_1\, ds_2,$$

where

$$h(s_1,s_2) = \langle(\alpha_{s_1}(x_1))(\alpha_{s_1+s_2}(x_2)),\phi\rangle$$

and

$$k(s_1,s_2) = \int f(s)f_1(s_1 - s)f_2(s_2 + s_1 - s)ds.$$

For a fixed $s_2$, note that $k(\cdot,s_2)$ is the convolution product $f * (f_1 \cdot f_{2,-s_2})$, where the dot denotes pointwise product and $g_t(s) = g(s-t)$; notice that $\hat{f}$ vanishes on $E + V + V$, while $(f_1 \cdot f_{2,-s_2})^\wedge$ (which is essentially the convolution of $\hat{f}_1$ and $\hat{f}_{2,-s_2}$) is supported inside spt $\hat{f}_1$ + spt $\hat{f}_{2,-s_2}$ which is contained in $E_1 + V + E_2 + V \subseteq E + V + V$. Consequently $\hat{f} \cdot (f_1 \cdot f_{2,-s_2})^\wedge \equiv 0$ and so, $k(\cdot,s_2) \equiv 0$. Yet another application of Fubini's theorem ensures, now, that $\langle\alpha(f)x,\phi\rangle = 0$. Since $\phi$ was arbitrary, this concludes the proof of case (i).

**Case (ii).** $x_1,x_2$ arbitrary.

For any $y$ in $M$, it follows from Ex. (3.2.5) (b) and Ex. (3.2.3) (d) that $y$ belongs to the $\sigma$-weak (and hence weak) closure of the set $\{\alpha(k)y: k \in \hat{C}, \|k\|_1 < 2\}$. This set is convex, and it is true that for convex subsets of $\mathcal{L}(\mathcal{H})$, the strong and weak closures coincide (cf. Ex. (3.2.11)). Select nets $\{f_i\}_{i\in I}$ and $\{g_j: j \in J\}$ such that $f_i,g_j \in \hat{C}$, $\|f_i\|_1$, $\|g_j\|_1 < 2$ and $\alpha(f_i)x_1 \to x_1$ and $\alpha(g_j)x_2 \to x_2$ strongly. Note that

$$\sup\{\|\alpha(f_i)x_1\|, \|\alpha(g_j)x_2\|\}_{\substack{i\in J \\ j\in J}} < \infty .$$

Since multiplication is jointly strongly continuous on bounded sets, $(\alpha(f_i)x_1)(\alpha(g_j)x_2) \to x_1x_2$ strongly, and hence weakly. By Case (i) and Prop. 3.2.9 (d), each of the products $(\alpha(f_i)x_1)(\alpha(g_j)x_2)$ belongs to $M(\alpha,E)$ which, by Lemma 3.2.8 (a), is $\sigma$-weakly closed; since, on bounded sets, the $\sigma$-weak and weak topologies coincide (cf. Ex. 0.3.1(b)), conclude that $x_1x_2 \in M(\alpha,E)$.

## Exercises

**(3.2.11)** Let $\phi$: $\mathcal{L}(\mathcal{H}) \to \mathbb{C}$ be a linear functional which is strongly continuous.

(a) Show that there exists an orthonormal set $\{\xi_1, ..., \xi_n\}$ in $\mathcal{H}$ and a constant $K > 0$ such that

$$|\phi(x)| \leqslant K\left[\sum_{i=1}^{n} \|x\xi_i\|^2\right]^{1/2}$$

for all $x$ in $\mathcal{L}(\mathcal{H})$. (Hint: First pick $\zeta_1, ..., \zeta_m$ in $\mathcal{H}$ and $\epsilon > 0$ such that $\|x\zeta_i\| < \epsilon$ for $1 \leqslant i \leqslant m \Rightarrow |\phi(x)| < 1$ (by continuity of $\phi$ at 0). If $\{\xi_1, ..., \xi_n\}$ is an orthonormal set such that $[\{\xi_1, ..., \xi_n\}] = [\{\zeta_1, ..., \zeta_m\}]$ then

$$\|x\zeta_i\| \leqslant \sum_{j=1}^{n} |\langle\zeta_i,\xi_j\rangle| \, \|x\xi_j\| \leqslant \|\zeta_i\|(\Sigma \, \|x\xi_j\|^2)^{1/2} .$$

Note that $\|x\xi_j\| = 0$ for all $j \Rightarrow \phi(x) = 0$. Show (by considering $K^{-1}x$) that $K = \epsilon^{-1}\max_j\|x\xi_j\|$ works.)

(b) There exists a (well-defined) bounded linear functional $\psi$ on $\tilde{\mathcal{H}}$, where

$$\tilde{\mathcal{H}} = \underbrace{\mathcal{H} \oplus \cdots \oplus \mathcal{H}}_{n \text{ copies}}$$

such that

$$\psi\left[\bigoplus_{j=1}^{n} x\xi_j\right] = \phi(x)$$

for all $x$ in $\mathcal{L}(\mathcal{H})$.

(c) There exist vectors $\eta_1, \ldots, \eta_n$ in $\mathcal{H}$ such that

$$\phi(x) = \sum_{j=1}^{n} \langle x\xi_j, \eta_j \rangle ;$$

in particular, $\phi$ is weakly continuous. (Hint: look at (b) and appeal to Riesz.)

(d) A convex subset of $\mathcal{L}(\mathcal{H})$ is weakly closed iff it is strongly closed. (Hint: the (locally convex version of the) Hahn-Banach theorem says that a closed convex set in a locally convex topological vector space can be separated from any point outside it by a continuous linear functional.) □

## 3.3. The Connes Spectrum of an Action

If $\alpha$ is an action of $G$ on $M$, we shall, as in Definition 2.5.13 (where only the case $G = \mathbb{R}$ was considered), denote by $M^\alpha$ the fixed point algebra: $M^\alpha = \{x \in M: \alpha_t(x) = x \ \forall t \in G\}$. Clearly $M^\alpha$ is a von Neumann subalgebra of $M$. For a projection $e$ in $M^\alpha$, $eMe$ may be viewed as a von Neumann algebra $M_e$ of operators on ran $e$. Since $e \in M^\alpha$, it follows that $\alpha$ induces an action $\alpha^e$ of $G$ on $M_e$ such that $\alpha_t^e(exe) = e(\alpha_t(x))e$. (The invariance $\alpha_t(e) = e$ is needed to ensure that this definition is unambiguous, and that $\alpha^e$ is an action. Verify this!)

**Proposition 3.3.1.** *Let $\alpha$ be an action of $G$ on $M$; let $e, e_1, e_2 \in P(M^\alpha)$ and let $E \subseteq \Gamma$ be closed.*

(a) $M_e(\alpha^e, E) = M(\alpha, E) \cap M_e$;

(b) $e_1 \leqslant e_2 \Rightarrow \mathrm{sp}\ \alpha^{e_1} \subseteq \mathrm{sp}\ \alpha^{e_2}$;

(c) *if $\mu \in M(G)$, $x \in M$ and $a,b \in M^\alpha$, then* $\alpha(\mu)(axb) = a(\alpha(\mu)x)b$;

(d) *if $x \in M$ and if $a$ and $b$ are invertible operators in $M^\alpha$, then* $\mathrm{sp}_\alpha(axb) = \mathrm{sp}_\alpha(x)$.

**Proof.** (a) If $x \in M_e$, note that $\alpha^e(f)x = \alpha(f)x$ for all $f$ in $L^1(G)$

$$\left[\int f(t)\alpha_t(exe)dt = \int f(t)e\ \alpha_t(x)e\ dt\right]$$

and hence

$$\mathrm{sp}_\alpha(x) = \mathrm{sp}_{\alpha^e}(x).$$

The desired assertion follows immediately.

(b) By Prop. 3.2.9 (b),

$$\mathrm{sp}\ \alpha^{e_1} = \left[\bigcup_{x\in M_{e_1}} \mathrm{sp}_{\alpha^{e_1}}(x)\right]^- = \left[\bigcup_{x\in M_{e_1}} \mathrm{sp}_\alpha(x)\right]^- \quad \text{(cf. (a))}$$

$$\subseteq \left[\bigcup_{x\in M_{e_2}} \mathrm{sp}_\alpha(x)\right]^- = \left[\bigcup_{x\in M_{e_2}} \mathrm{sp}_{\alpha^{e_2}}(x)\right]^- = \mathrm{sp}\ \alpha^{e_2}.$$

(c) For any $\phi$ in $M_*$,

$$\begin{aligned} \langle\alpha(\mu)(axb),\phi\rangle &= \int \langle\alpha_t(axb),\phi\rangle d\mu(t) \\ &= \int \langle a\ \alpha_t(x)b,\phi\rangle d\mu(t) \\ &= \langle a(\alpha(\mu)x)b,\phi\rangle\ , \end{aligned}$$

since multiplication is separately continuous in the $\sigma$-weak topology (cf. Ex. (0.3.3) (d)).

(d). Conclude from Prop. 3.2.10 and Lemma 3.2.8(c) that

$$\begin{aligned} \mathrm{sp}_\alpha(axb) &\subseteq (\mathrm{sp}_\alpha(a) + \mathrm{sp}_\alpha(x) + \mathrm{sp}_\alpha(b))^- \\ &\subseteq (\{0\} + \mathrm{sp}_\alpha(x) + \{0\})^- \\ &= \mathrm{sp}_\alpha(x); \end{aligned}$$

the reverse inclusion follows, since

$$\mathrm{sp}_\alpha(x) = \mathrm{sp}_\alpha(a^{-1}(axb)b^{-1}) \subseteq \mathrm{sp}_\alpha(axb). \qquad \square$$

**Definition 3.3.2.** If $\alpha$ is an action of $G$ on $M$, the Connes spectrum of $\alpha$, denoted by $\Gamma(\alpha)$, is defined thus: $\Gamma(\alpha) = \cap\{\mathrm{sp}\ \alpha^e\colon 0 \neq e \in \mathcal{P}(M^\alpha)\}$. $\square$

It is clear from the definition that $\Gamma(\alpha)$ is always a closed subset of $\Gamma$. The content of the following assertion is that in defining $\Gamma(\alpha)$, it suffices to consider only non-zero projections which are central in $M^\alpha$.

**Proposition 3.3.3.** *For an action $\alpha$ of $G$ on $M$,*

$$\Gamma(\alpha) = \cap \{\text{sp } \alpha^e: 0 \neq e \in P(Z(M^\alpha))\};$$

*in particular, if $M^\alpha$ is a factor, then* $\Gamma(\alpha) = \text{sp } \alpha$.

**Proof.** We shall show that if $0 \neq e \in P(M^\alpha)$, then there exists a non-zero $\bar{e}$ in $P(Z(M^\alpha))$ such that $\text{sp } \alpha^e = \text{sp } \alpha^{\bar{e}}$. In fact, define $\bar{e} = \bigvee\{ueu^*: u \in U(M^\alpha)\}$. On the one hand, $\bar{e} \in M^\alpha$ since it is the supremum of a family of projections in $M^\alpha$; on the other hand, it is clear that $u\bar{e}u^* = \bar{e}$ for all $u$ in $U(M^\alpha)$ and so $\bar{e} \in (M^\alpha)'$ (by Scholium 0.4.8); thus $\bar{e} \in P(Z(M^\alpha))$.

In order to prove $\text{sp } \alpha^e = \text{sp } \alpha^{\bar{e}}$, it suffices (by Prop. 3.2.9 (e) and Prop. 3.3.1 (a)) to show that, for any closed set $E$ in $\Gamma$, $M(\alpha,E) \cap M_e \neq \{0\}$ if and only if $M(\alpha,E) \cap M_{\bar{e}} \neq \{0\}$. Since $M_e \subseteq M_{\bar{e}}$ (as $e \leq \bar{e}$), the "only if" part is clear; conversely suppose $0 \neq x \in M(\alpha,E) \cap M_{\bar{e}}$; since $x = \bar{e}x\bar{e}$, there exist $u,v \in U(M^\alpha)$ such that $(ueu^*)x(vev^*) \neq 0$; then $y = eu^*xve \neq 0$, and it is clear that $y \in M_e$; and since $e$, $u$, $v \in M^\alpha$, it follows from Prop. 3.2.10 and Lemma 3.2.8 that $\text{sp}_{\alpha^e}(y) \subseteq \text{sp}_\alpha(x) \subseteq E$, whence $M(\alpha,E) \cap M_e \neq (0)$, as desired. □

We shall head towards the main result of this section via a sequence of lemmas.

**Lemma 3.3.4.** *Let $\{V_j: j \in \Lambda\}$ be an open cover of $\Gamma$ and let $x \in M$. If $x \neq 0$, there exists $f$ in $\hat{C}$ such that* $\text{spt } \hat{f} \subseteq V_j$ *for some $j$ in $\Lambda$ and* $\alpha(f)x \neq 0$.

**Proof.** Let $I_0$ be the set of linear combinations of elements of $L^1(G)$ whose Fourier transforms are supported inside compact subsets of members of the cover $\{V_j\}$. It is clear that $I_0$ is an ideal in $L^1(G)$; hence, the closure $I$ of $I_0$ in $L^1(G)$ is a closed ideal in $L^1(G)$. If $\gamma \in \Gamma$, pick $j \in \Lambda$ such that $\gamma \in V_j$; then, using Prop. 3.2.4 (a), choose $f$ in $\hat{C}$ such that $\hat{f}(\gamma) = 1$ and $\text{spt } \hat{f} \subseteq V_j$; thus, for each $\gamma$ in $\Gamma$, there exists an $f$ in $I$ such that $\hat{f}(\gamma) \neq 0$. On the other hand, it is a fact (cf. [Loo], Section 37) that if $I$ is a closed ideal in $L^1(G)$ and if $I \neq L^1(G)$, then there exists a $\gamma$ in $\Gamma$ such that $\hat{f}(\gamma) = 0$ for all $f$ in $I$. Conclusion: $I_0$ is dense in $L^1(G)$.

This conclusion, together with Ex. (3.2.3) (a) and the fact that $\|\alpha(g)\| \leq \|g\|_1$ for all $g$ in $L^1(G)$, completes the proof of the lemma. □

**Lemma 3.3.5.** *If $e_1$ and $e_2$ are non-zero projections in $M^\alpha$, which are equivalent relative to $M$ (as in Def.* 1.1.1), *then*

$$\Gamma(\alpha^{e_1}) = \Gamma(\alpha^{e_2}).$$

**Proof.** It suffices (by symmetry) to show that if $\gamma \in \Gamma(\alpha^{e_1})$, if $V$ is any neighborhood of $\gamma$, and if $f_2$ is any non-zero projection in $M^\alpha$ such that $f_2 \leq e_2$, then

$$M(\alpha,V) \cap M_{f_2} \neq \{0\}.$$

Let $U$ and $W$ be neighborhoods of $\gamma$ and 0 (in $\Gamma$) respectively such that $\overline{U+W} \subseteq V$. Next, choose an open cover $\{V_j\}$ of $\Gamma$ such that $V_j - V_j \subseteq W$ for all $j$.

By hypothesis, there is a partial isometry $u$ in $M$ such that $u^*u = e_1$ and $uu^* = e_2$. Since $f_2u \neq 0$, use Lemma 3.3.4 to choose a $g$ in $\hat{C}$ such that spt $\hat{g}$ − spt $\hat{g} \subseteq W$ and $\alpha(g)(f_2u) \neq 0$. Let $x = \alpha(g)(f_2u)$ and note that (by Prop. 3.3.1 (c)) $x = f_2xe_1$, since $f_2u = f_2 \cdot f_2u \cdot e_1$, and that $\mathrm{sp}_\alpha(x) \subseteq$ spt $\hat{g}$ (by Prop. 3.2.9 (d)) so that $\mathrm{sp}_\alpha(x) - \mathrm{sp}_\alpha(x) \subseteq W$.

Let $f_1 = \vee\{rp\ \alpha_t(x^*)\colon t \in G\}$. It is clear that $0 \neq f_1 \in P(M^\alpha)$; also, as $\alpha_t(x) = \alpha_t(f_2xe_1) = f_2\alpha_t(x)e_1$, note that $f_1 \leqslant e_1$. Since $U$ is a neighborhood of $\gamma$ and $\gamma \in \Gamma(\alpha^{e_1})$, there exists a non-zero element $y$ in $M(\alpha,U) \cap M_{f_1}$. The definition of $f_1$ and the equality $y = f_1y\,f_1$ ensure the existence of $t_1$, $t_2$ in $G$ such that

$$z = \alpha_{t_1}(x)y\ \alpha_{t_2}(x)^* \neq 0.$$

Since $\alpha_t(x) = f_2\ \alpha_t(x)$ for all $t$, it is clear that $z = f_2z\ f_2 \in M_{f_2}$, and that

$$\begin{aligned}\mathrm{sp}_\alpha(z) &\subseteq (\mathrm{sp}_\alpha(\alpha_{t_1}(x)) + \mathrm{sp}_\alpha(y) - \mathrm{sp}_\alpha(\alpha_{t_2}(x)))^- \\ &= (\mathrm{sp}_\alpha(x) + \mathrm{sp}_\alpha(y) - \mathrm{sp}_\alpha(x))^- \\ &\subseteq (W + U)^- \\ &\subseteq V,\end{aligned}$$

and hence,

$$0 \neq z \in M(\alpha,V) \cap M_{f_2}. \qquad \square$$

As in Definition 3.1.2 (b) -- where we restricted ourselves to the case $G = \mathbb{R}$ -- we shall say that actions $\alpha$ and $\beta$ of $G$ on $M$ are outer equivalent ($\alpha \overset{\circ}{\sim} \beta$, in symbols) if there is a strongly continuous map $t \to u_t$ from $G$ into $U(M)$ such that $u_{t+s} = u_t\alpha_t(u_s)$ and $\beta_t(x) = u_t\alpha_t(x)u_t^*$, for all $s,t$ in $G$ and $x$ in $M$.

**Lemma 3.3.6.** *Suppose $\alpha \overset{\circ}{\sim} \beta$, where $\alpha$ and $\beta$ are actions of $G$ on $M$. Then there exists an action $\gamma$ of $G$ on $\tilde{M} = M \otimes M_2(\mathbb{C})$ (cf. proof of Theorem 3.1.1) such that*

$$\gamma_t(x \otimes e_{11}) = \alpha_t(x) \otimes e_{11}$$

*and*

$$\gamma_t(x \otimes e_{22}) = \beta_t(x) \otimes e_{22}$$

*for all t in G and x in M.*

**Proof.** If $\{u_t\}$: $\{\alpha_t\} \overset{\circ}{\sim} \{\beta_t\}$, simply define

$$\gamma_t\left(\begin{bmatrix} x_{11} & x_{12} \\ x_{21} & x_{22} \end{bmatrix}\right) = \begin{bmatrix} \alpha_t(x_{11}) & \alpha_t(x_{12})u_t^* \\ u_t\alpha_t(x_{21}) & \beta_t(x_{22}) \end{bmatrix},$$

and verify that $\gamma$ does the job. □

**Theorem 3.3.7.** *Let $\alpha$ be an action of G on M.*

(a) $\Gamma(\alpha) + \text{sp}\,\alpha = \text{sp}\,\alpha$;
(b) $\Gamma(\alpha)$ *is a closed subgroup of* $\Gamma$;
(c) *If $\beta$ is another action of G on M, which is outer equivalent to $\alpha$, then* $\Gamma(\alpha) = \Gamma(\beta)$.

**Proof.** (a) If $0 \neq e \in P(M^\alpha)$, then $e$ is an identity for $M_e$ and hence $\text{sp}_{\alpha^e}(e) = \{0\}$ (cf. Lemma 3.2.8 (c)), so that

$$0 \in \text{sp}_{\alpha^e}(e) \subseteq \text{sp}\,\alpha^e;$$

so $0 \in \Gamma(\alpha)$ and consequently $\Gamma(\alpha) + \text{sp}\,\alpha \supseteq \text{sp}\,\alpha$. Suppose, conversely, that $\gamma_1 \in \Gamma(\alpha)$ and $\gamma_2 \in \text{sp}\,\alpha$; we need to verify that $M(\alpha,V) \neq \{0\}$ for any neighborhood $V$ of $(\gamma_1 + \gamma_2)$. Pick neighborhoods $V_i$ of $\gamma_i$ such that $\overline{V_1 + V_2} \subseteq V$. Since $\gamma_2 \in \text{sp}\,\alpha$, there exists a non-zero element $x_2$ in $M(\alpha,V_2)$. Put $e = \vee\{\text{rp}\,\alpha_t(x_2^*)\colon t \in G\}$ and notice (as in the proof of Lemma 3.3.5) that $0 \neq e \in P(M^\alpha)$. Since $\gamma_1 \in \Gamma(\alpha)$, there exists a non-zero element $x_1$ in $M(\alpha,V_1) \cap M_e$. It follows from the definition of $e$ and the fact that $0 \neq x_1 = ex_1$, that there exists $t$ in $G$ such that $\alpha_t(x_2)x_1 \neq 0$; set $x = \alpha_t(x_2)x_1$ and notice that $x \in M(\alpha,V)$ (as

$$\begin{aligned}\text{sp}_\alpha(x) &\subseteq (\text{sp}_\alpha(\alpha_t(x_2)) + \text{sp}_\alpha(x_1))^- = (\text{sp}_\alpha(x_2) + \text{sp}_\alpha(x_1))^- \\ &\subseteq (V_2 + V_1)^- \subseteq V).\end{aligned}$$

(b) Fix a non-zero $e$ in $P(M^\alpha)$ and conclude from (a) that $\Gamma(\alpha^e) + \text{sp}\,\alpha^e = \text{sp}\,\alpha^e$. Since clearly $\Gamma(\alpha) \subseteq \Gamma(\alpha^e) \subseteq \text{sp}\,\alpha^e$, infer that $\Gamma(\alpha) + \Gamma(\alpha) \subseteq \text{sp}\,\alpha^e$; allow $e$ to vary and conclude that $\Gamma(\alpha) + \Gamma(\alpha) \subseteq \Gamma(\alpha)$.

Also, infer from Prop. 3.2.9 (a), (b) that $\text{sp}\,\alpha^e = -\text{sp}\,\alpha^e$, for every non-zero $e$ in $P(M^\alpha)$. Hence $\Gamma(\alpha) = -\Gamma(\alpha)$. Since we have already observed in the proof of (a) that $0 \in \Gamma(\alpha)$, we have now shown that $\Gamma(\alpha)$ is a subgroup of $\Gamma$. The fact that $\Gamma(\alpha)$ is a closed set is obvious.

(c) Let $\tilde{M} = M \otimes M_2(\mathbb{C})$ and $\gamma$ be as in Lemma 3.3.6. Then $1 \otimes e_{11}$ and $1 \otimes e_{22}$ are readily verified to be non-zero projections in $\tilde{M}^\gamma$, and $1 \otimes e_{11} = \tilde{u}^*\tilde{u}$ and $1 \otimes e_{22} = \tilde{u}\tilde{u}^*$ where $\tilde{u} = 1 \otimes e_{21}$; so, by Lemma

3.3.5, conclude that

$$\Gamma(\gamma^{1\otimes e_{11}}) = \Gamma(\gamma^{1\otimes e_{22}}).$$

On the other hand, it is clear that there are natural isomorphisms of dynamical systems:

$$(M,G,\alpha) \cong (\tilde{M}_{1\otimes e_{11}}, G, \gamma^{1\otimes e_{11}})$$

and

$$(M,G,\beta) \cong (\tilde{M}_{1\otimes e_{22}}, G, \gamma^{1\otimes e_{22}});$$

hence $\Gamma(\alpha) = \Gamma(\beta)$. □

We are finally in a position to harvest the consequences of the elaborate machinery we have developed in the last two sections. Let $G = \mathbb{R}$. We shall identify the dual group $\Gamma$ with the multiplicative group $\mathbb{R}_+^*$ of positive real numbers, the duality being given by $\langle t,\gamma\rangle = \gamma^{it}$. If $M$ is an arbitrary von Neumann algebra, it follows from Theorems 3.1.1 and 3.3.7 (c) that $\Gamma(\sigma^\phi) = \Gamma(\sigma^\psi)$, where $\sigma^\phi$ and $\sigma^\psi$ are the "modular flows" associated with fns weights $\phi$ and $\psi$ on $M$.

**Definition 3.3.8.** For an arbitrary von Neumann algebra $M$, define $\Gamma(M) = \Gamma(\sigma^\phi)$, where $\phi$ is any fns weight on $M$. □

**Corollary 3.3.9.** (a) $\Gamma(M)$ *is one and only one of the following sets*:

(0) $\Gamma(M) = \{1\}$;
($\lambda$) $\Gamma(M) = \{\lambda^n : n \in \mathbb{Z}\}$ *for some* $\lambda$ *in* $(0,1)$;
(1) $\Gamma(M) = (0,\infty)$.

(b) *If* $M$ *is semifinite, then* $\Gamma(M) = \{1\}$.

**Proof.** (a) This is a consequence of the fact that the only non-trivial closed subgroups of $\mathbb{R}$ (and consequently, of the isomorphic group $\mathbb{R}_+^*$) are the cyclic ones.

(b) If $\tau$ is a fns trace on $M$, then $\sigma_t^\tau(x) = x$ for all $t$ in $\mathbb{R}$ and $x$ in $M$; so $M = M^\tau$ and (cf. Lemma 3.2.8 (c)) $\Gamma(M) = \Gamma(\sigma^\tau) = \{1\}$. □

**Definition 3.3.10.** A factor $M$ is said to be of type

(0) $\mathrm{III}_0$, if $M$ is of type III and $\Gamma(M) = \{1\}$;
($\lambda$) $\mathrm{III}_\lambda$, if $\Gamma(M) = \{\lambda^n : n \in \mathbb{Z}\}$ (for $\lambda$ in $(0,1)$);
(1) $\mathrm{III}_1$, if $\Gamma(M) = (0,\infty)$. □

It follows from Corollary 3.3.9 (b) that factors of type $\mathrm{III}_\lambda$ are of type III, for $0 \leq \lambda \leq 1$. Examples of factors of all these types will be exhibited in §4.3. The existence of factors of type $\mathrm{III}_0$ shows that the condition $\Gamma(M) = \{1\}$, while being necessary for semifiniteness, is by no means sufficient.

## 3.4. Alternative Descriptions of $\Gamma(M)$

As the title of this section announces, we shall derive some equivalent descriptions of $\Gamma(M)$, which will be useful when it comes to explicit computation.

**Lemma 3.4.1.** *Let* $\alpha$ *be an action of* $G$ *on a factor* $M$ *of type* III. *Then,*

$$\Gamma(\alpha) = \cap \{\mathrm{sp}\ \beta : \alpha \overset{\circ}{\sim} \beta\}.$$

**Proof.** If $\alpha \overset{\circ}{\sim} \beta$, then $\Gamma(\alpha) = \Gamma(\beta) \subseteq \mathrm{sp}\ \beta$ and so $\Gamma(\alpha) \subseteq \cap\{\mathrm{sp}\ \beta: \alpha \overset{\circ}{\sim} \beta\}$. The reverse inclusion will be proved once we show that for each non-zero $e$ in $P(M^\alpha)$, there is an action $\beta$ such that $\alpha \overset{\circ}{\sim} \beta$ and $\mathrm{sp}\ \alpha^e = \mathrm{sp}\ \beta$.

So, suppose $0 \neq e \in P(M^\alpha)$; since $M$ is of type III, any two non-zero projections in $M$ are equivalent, and in particular, there exists an isometry $u$ in $M$ such that $u^*u = 1$ and $uu^* = e$. Since $e \in M^\alpha$, it follows that $\alpha_t(u)^*\alpha_t(u) = 1$ and $\alpha_t(u)\alpha_t(u)^* = e$ for each $t$ in $G$; so the equation $v_t = u^*\alpha_t(u)$ defines a strongly continuous unitary path in $M$. Notice that for any $s,t$ in $G$,

$$v_t\alpha_t(v_s) = u^*\alpha_t(u)\alpha_t(u^*\alpha_s(u)) = u^*e\alpha_{t+s}(u) = v_{t+s}$$

(since $u^*e = u^*$); hence, by Ex. (3.1.3) (b), the equation $\beta_t(x) = v_t\alpha_t(x)v_t^*$ defines an action $\beta$ of $G$ on $M$, which is outer equivalent to $\alpha$.

Observe now that $\beta_t(x) = u^*\alpha_t(ux\ u^*)u$; the operator $u$ may be regarded as a unitary operator from $\mathcal{H}$ to ran $e$, which sets up an isomorphism between the dynamical system $(M,G,\beta)$ and $(M_e,G,\alpha^e)$ -- explicitly, if $\pi(x) = uxu^*$, then $\pi$: $M \to M_e$ is a von Neumann algebra isomorphism such that $\alpha_t^e \circ \pi = \pi \circ \beta_t$ for all $t$ in $G$. Consequently, $\mathrm{sp}\ \beta = \mathrm{sp}\ \alpha^e$ and the proof is complete. □

The next lemma is a sort of converse to the unitary cocycle theorem. Since the techniques involved in the proof are not strictly in keeping with those treated here, we shall simply state the result without proof; the reader desirous of seeing a proof of the result is directed to [Kal 1].

**Lemma 3.4.2.** *Let* $\phi$ *be a* fns *weight on a factor* $M$. *If* $\beta$ *is any action of* $\mathbb{R}$ *on* $M$ *such that* $\sigma^\phi \overset{\circ}{\sim} \beta$, *there exists a unique* fns *weight* $\psi$ *on* $M$ *such that* $\beta_t = \sigma_t^\psi$ *for all* $t$ *in* $\mathbb{R}$. □

**Lemma 3.4.3.** *If* $M$ *is a factor, then*

$$\Gamma(M) = \cap \{\mathrm{sp}\ \sigma^{\phi}: \phi \text{ a fns weight on } M\}.$$

**Proof.** If $M$ is semifinite, the assertion follows from Corollary 3.3.9 (b) and the obvious fact that sp $\sigma^{\tau} = \{1\}$ if $\tau$ is a fns trace on $M$. If $M$ is of type III, the assertion is a consequence of Theorem 3.1.1 and Lemmas 3.4.1 and 3.4.2. □

The natural next step of examining sp $\sigma^{\phi}$ is taken up in the following lemma.

**Lemma 3.4.4.** *If $\phi$ is a* fns *weight on $M$, then* sp $\sigma^{\phi}$ = sp $\Delta_{\phi} \cap (0,\infty)$. *(The symbol* sp $\Delta_{\phi}$ *denotes, of course, the spectrum of the positive self-adjoint operator $\Delta_{\phi}$; also, as at the end of Section* 4.3, *we are identifying the dual group of $\mathbb{R}$ with $\mathbb{R}^*_+$.)*

**Proof.** If $f \in L^1(\mathbb{R})$, then $\hat{f}$ is the bounded continuous function defined on $(0,\infty)$ by $\hat{f}(\lambda) = \int f(t)\lambda^{-it}dt$; since $\Delta_{\phi}$ is invertible, it follows that (though $\hat{f}$ is not defined at 0) $\hat{f}(\Delta_{\phi}^{-1})$ is a meaningfully defined bounded operator on $\mathcal{H}_{\phi}$, and that

$$\hat{f}(\Delta_{\phi}^{-1}) = \int_{(0,\infty)} f(t)\ \Delta_{\phi}^{it}\ dt$$

the integral being interpreted weakly.

**Assertion:** $x \in N_{\phi} \Rightarrow y = \sigma^{\phi}(f)x \in N_{\phi}$ and $\eta_{\phi}(\sigma^{\phi}(f)x) = \hat{f}(\Delta_{\phi}^{-1})\eta_{\phi}(x)$.

In case $\phi$ is finite, the first assertion is trivial (since then, $N_{\phi} = M$), while

$$\eta_{\phi}(\sigma^{\phi}(f)x) = (\sigma^{\phi}(f)x)\Omega_{\phi} = \int f(t)\sigma_t^{\phi}(x)\Omega\ dt = \hat{f}(\Delta_{\phi}^{-1})x\Omega_{\phi},$$

since

$$\sigma_t^{\phi}(x)\Omega_{\phi} = \Delta_{\phi}^{it}x\Delta_{\phi}^{-it}\Omega_{\phi} = \Delta_{\phi}^{it}x\Omega_{\phi}.$$

If $\phi$ is infinite, assume (with no loss of generality) that $f \geq 0$ and that $\int f(t)dt = 1$; notice that the set

$$K = \{z \in N_{\phi}: \|z\| \leq \|x\|,\ \|\eta_{\phi}(z)\| \leq \|\eta_{\phi}(x)\|\}$$

is a convex, norm-bounded set. The set $K$ is $\sigma$-strongly* closed since multiplication is jointly strongly continuous on bounded sets and since $\phi$, being normal, is $\sigma$-weakly lower semicontinuous (cf. Prop. 2.4.9). It is a fact that a convex set is $\sigma$-strongly* closed if and only if it is $\sigma$-weakly closed. (The reader is invited to fill in the details: one must show, using an argument similar to the one developed in Ex. (3.2.11), that any $\sigma$-strongly* continuous linear functional on $\mathcal{L}(\mathcal{H})$ is $\sigma$-weakly continuous. With the notation as in Ex. (3.2.11), one must consider

$$\tilde{\mathcal{H}} = \bigoplus_{\substack{n=-\infty \\ n\neq 0}}^{\infty} \mathcal{H}_n, \text{ with } \mathcal{H}_n = \mathcal{H} \text{ for } n > 0, \text{ and } \mathcal{H}_n = \overline{\mathcal{H}}$$

(the conjugate Hilbert space) for $n < 0$.)

Thus $K$ is a $\sigma$-weakly closed convex set which contains $\sigma_t^\phi(x)$ for all $t$; since $f \geqslant 0$ and $\int f(t)dt = 1$, conclude that

$$y = \int f(t)\ \sigma_t^\phi(x)dt \in K;$$

observe that for arbitrary $z$ in $N_\phi$,

$$\begin{aligned}
<\eta_\phi(y),\eta_\phi(z)> &= \phi(z^*y) \\
&= \int f(t)<z^*\sigma_t^\phi(x),\phi>dt \\
&= \int f(t)<\Delta_\phi^{it}\ \eta_\phi(x),\ \eta_\phi(z)>dt \\
&= <\hat{f}(\Delta_\phi^{-1})\eta_\phi(x),\ \eta_\phi(z)>\ ,
\end{aligned}$$

thus establishing the validity of the assertion.

Hence, if $f \in L^1(\mathbb{R})$,

$$\begin{aligned}
\sigma^\phi(f) = 0 &\Leftrightarrow \sigma^\phi(f)x = 0 \quad \forall x \text{ in } N_\phi \\
&\Leftrightarrow \hat{f}(\Delta_\phi^{-1})\eta_\phi(x) = 0 \quad \forall x \text{ in } N_\phi \\
&\Leftrightarrow \hat{f}(\Delta_\phi^{-1}) = 0 \\
&\Leftrightarrow \hat{f} \text{ vanishes on } (\text{sp } \Delta_\phi^{-1} \cap \mathbb{R}_+^*) \\
&\Leftrightarrow \hat{f} \text{ vanishes on } (\text{sp } \Delta_\phi \cap \mathbb{R}_+^*),
\end{aligned}$$

the last equivalence following from the equation $J_\phi\ \Delta_\phi\ J_\phi = \Delta_\phi^{-1}$. Set $I = \{f \in L^1(\mathbb{R}): \sigma^\phi(f) = 0\}$ and $E = \text{sp } \Delta_\phi \cap \mathbb{R}_+^*$; the above statement translates into $I = I(E)$, and hence, sp $\sigma^\phi = I^\perp = I(E)^\perp = E$, as desired. □

**Theorem 3.4.5.** *If $M$ is a factor, then*

$$\Gamma(M) = \mathbb{R}_+^* \cap (\cap\{\text{sp } \Delta_\phi\text{: } \phi \text{ a fns weight on } M\}).$$

**Proof.** This is an immediate consequence of Lemmas 3.4.3 and 3.4.4. □

Following Connes, let us introduce the notation

$$S(M) = \cap\{\text{sp } \Delta_\phi\text{: } \phi \text{ a fns weight on } M\}.$$

Then $S(M)$ is clearly an isomorphism-invariant; further, if $M$ is a factor, it follows, from Theorems 3.3.7 and 3.4.5, that $S(M)$ is a closed subset of $[0,\infty)$ such that sp $\Delta_\phi$ is left invariant under multiplication by any positive number in $S(M)$ for any fns weight $\phi$ on $M$. In terms of $S(M)$, Theorem 3.4.5 becomes:

$$\Gamma(M) = S(M) \cap \mathbb{R}_+^*, \quad \text{if } M \text{ is a factor.}$$

The sets $\Gamma(M)$ and $S(M)$ can thus differ at most by the number 0. The content of the following proposition is that there is no difference precisely when $M$ is semifinite.

**Proposition 3.4.6.** *The following conditions on a factor $M$ are equivalent:*

(i) *$M$ is semifinite;*
(ii) $S(M) = \{1\}$,
(iii) $0 \notin S(M)$.

**Proof.** (i) $\Rightarrow$ (ii): Since $1 \in \Gamma(M)$ (cf. Th. 3.3.7), it follows from Theorem 3.4.5 that $1 \in S(M)$. Conversely, if $\tau$ is a fns trace on $M$, then $\Delta_\tau = 1_{\mathcal{H}_t}$ and so $S(M) \subseteq \text{sp } \Delta_\tau = \{1\}$.

(ii) $\Rightarrow$ (iii): $0 \neq 1$.

(iii) $\Rightarrow$ (i): By definition of $S(M)$, the assumption (iii) is that there exists a fns weight $\phi$ on $M$ such that $0 \notin \text{sp } \Delta_\phi$. Since $\Delta_\phi = J_\phi \Delta_\phi^{-1} J_\phi$, conclude that sp $\Delta_\phi \subseteq (\epsilon,\epsilon^{-1})$ for some $\epsilon$ in $(0,1)$. So, $H = \log \Delta_\phi$ is a bounded operator on $\mathcal{H}_\phi$ such that

$$e^{itH}\pi_\phi(M)e^{-itH} = \pi_\phi(M) \qquad \text{for all } t \text{ in } \mathbb{R}\ .$$

A result due to Sakai (cf. [Sak]) -- whose proof relies on some facts concerning derivations of von Neumann algebras -- states that under precisely these conditions, there must exist a strongly continuous one-parameter unitary group $\{u_t\}$ in $\pi_\phi(M)$ such that $e^{itH}xe^{-itH} = u_t x u_t^*$ for all $x$ in $\pi_\phi(M)$ and $t$ in $\mathbb{R}$; in other words, the flow $\sigma^\phi$ is inner, and hence, by Theorem 3.1.6, $M$ is semifinite. □

Since $\Delta_\phi$ and $\Delta_\phi^{-1}$ are anti-unitarily equivalent, the above result shows that if $M$ is a type III factor, then $\Delta_\phi$ is necessarily unbounded, for every fns weight $\phi$ on $M$.

Notice also that, as a consequence of Theorem 3.4.5 and Proposition 3.4.6, a factor $M$ is of type $\text{III}_0$, $\text{III}_\lambda$ $(0 < \lambda < 1)$ or $\text{III}_1$ according as $S(M)$ is $\{0,1\}$, $\{0\} \cup \{\lambda^n : n \in \mathbb{Z}\}$ or $[0,\infty)$.

We shall conclude this section with another description of $\Gamma(M)$ that has the advantage (over that given by Theorem 3.4.5) that it is in terms of any one fixed fns weight on $M$.

**Proposition 3.4.7.** *Let $\phi$ be a* fns *weight on a factor $M$. If $0 \neq e \in P(M^\phi)$, let $\phi_e = \phi|M_{e,+}$.*

(a) $\phi_e$ *is a* fns *weight on* $M_e$;
(b) $\Gamma(M) = \mathbb{R}_+^* \cap \bigcap\{\text{sp } \Delta_{\phi_e};\ 0 \neq e \in P(Z(M^\phi))\}$; *in particular, if* $M^\phi$ *is a factor, then* $\Gamma(M) = \mathbb{R}_+^* \cap \text{sp } \Delta_\phi$.

**Proof.** (a) It is clear that $\phi_e$ is a faithful and normal weight on $M_e$. Since $e \in M^\phi$, it follows from Theorem 2.5.14' that $e\mathfrak{D}_\phi e \subseteq \mathfrak{D}_\phi$; so

$$\mathfrak{D}_{\phi_e} \supseteq e\mathfrak{D}_\phi e.$$

The semifiniteness of $\phi$ ensures the existence of a monotone net $\{x_i\}$ in $\mathfrak{D}_\phi$ such that $x_i \nearrow 1$ (cf. Ex. (2.4.8)); then $\{ex_ie\}$ is a monotone net in $\mathfrak{D}_{\phi_e}$ which converges weakly to $e$, the identity of $M_e$; consequently $\phi_e$ is semifinite.

(b) In view of Prop. 3.3.3 and Lemma 3.4.4, it would suffice to show that $(\sigma^\phi)^e = \sigma^{\phi_e}$ for non-zero $e$ in $P(Z(M^\phi))$. Since $N_{\phi_e} \subseteq N_\phi$, it is trivial to verify that $\phi_e$ satisfies the KMS condition with respect to the flow $(\sigma^\phi)^e$, and the conclusion follows. □

**Corollary 3.4.8.** *If M is a factor, then*

$$S(M) = \bigcap\{\text{sp } \Delta_{\phi_e} : 0 \neq e \in P(Z(M^\phi))\},$$

*for any* fns *weight* $\phi$ *on M, with* $\phi_e$ *as in Proposition* 3.4.7.

**Proof. Case (i)**: $M$ is of type III.

In this case, $0 \in S(M)$, by Prop. 3.4.6. If $0 \neq e \in P(Z(M^\phi))$, since $M$ is of type III, there exists an isometry $u$ in $M$ such that $u^*u = 1$ and $uu^* = e$; the map $x \to uxu^*$ is a von Neumann algebra isomorphism of $M$ onto $M_e$ and hence $M_e$ is also a factor of type III; so, by Prop. 3.4.6 and Prop. 3.4.7 (a), $0 \in \text{sp } \Delta_{\phi_e}$.

**Case (ii)**: $M$ is semifinite.

In this case, $0 \notin S(M)$, by Prop. 3.4.6. We must exhibit a non-zero $e$ in $P(Z(M^\phi))$ such that $0 \notin \text{sp } \Delta_{\phi_e}$, or, equivalently, such that $\Delta_{\phi_e}$ is bounded. Let $\tau$ be a fns trace on $M$. So, by Theorem 2.6.3, there exists an invertible positive self-adjoint operator $H \eta M$ such that $\phi = \tau(H\cdot)$. Pick $\epsilon > 0$ such that $e = 1_{(\epsilon,1/\epsilon)}(H) \neq 0$. We know -- by Theorem 3.1.10 -- that $\sigma_t^\phi(x) = H^{it}xH^{-it}$, for $x$ in $M$ and $t$ in $\mathbb{R}$; thus $x \in M^\phi$ if and only if $x$ commutes with $1_E(H)$ for all Borel sets $E$; in particular, $e \in P(Z(M^\phi))$. It follows from $\epsilon e \leq He \leq \epsilon^{-1}e$ that

$$y \in M_{e,+} \Rightarrow \epsilon\tau(y) \leq \phi(y) \leq \epsilon^{-1}\tau(y);$$

so,

$$\begin{aligned} x \in M_e \Rightarrow \phi(xx^*) &\leqslant \epsilon^{-1}\tau(xx^*) \\ &= \epsilon^{-1}\tau(x^*x) \\ &\leqslant \epsilon^{-2}\phi(x^*x), \end{aligned}$$

and hence,

$$\begin{aligned} x \in N_{\phi_e} \cap N_{\phi_e}^* \Rightarrow \|\Delta_{\phi_e}^{1/2}\eta_{\phi_e}(x)\|^2 &= \|J_{\phi_e}\Delta_{\phi_e}^{1/2}\eta_{\phi_e}(x)\|^2 \\ &= \|\eta_{\phi_e}(x^*)\|^2 \\ &= \phi(xx^*) \\ &\leqslant \epsilon^{-2}\phi(x^*x) \\ &= \epsilon^{-2}\|\eta_{\phi_e}(x)\|^2, \end{aligned}$$

so that $\|\Delta_{\phi_e}^{1/2}\| \leqslant \epsilon^{-1} < \infty$, as desired. □

**Exercises**

**(3.4.9)** For any factor $M$, show that

(a) $\Gamma(M) = \mathbb{R}_+^* \cap \cap\{\text{sp } \Delta_{\phi_e} : 0 \neq e \in P(M^\phi)\}$;

(b) $S(M) = \cap\{\text{sp } \Delta_{\phi_e} : 0 \neq e \in P(M^\phi)\}$.

(Hint: Imitate the proofs of Prop. 3.4.7 (b) and Corollary 3.4.8.)

# Chapter 4
# CROSSED-PRODUCTS

The crossed-product construction was first employed by Murray and von Neumann to exhibit examples of factors of types I, II and III. The set-up is as follows: one starts with a dynamical system $(M,G,\alpha)$ -- with $G$ not necessarily abelian -- and constructs an associated von Neumann algebra $\underset{\sim}{M}$ (usually denoted by $M \otimes_\alpha G$) on a larger Hilbert space $\underset{\sim}{\mathcal{H}}$.

Section 4.1 discusses this construction when $G$ is a countable discrete group, and develops some of the features of the crossed product; for instance, a necessary and sufficient condition for $\underset{\sim}{M}$ to be a factor, is given in terms of the action $\alpha$.

In Section 4.2, we assume that $M$ is semifinite and use a fns trace on $M$ to construct a fns weight $\phi$ on $\underset{\sim}{M}$, whose associated modular operator is explicitly computed; this description is used to compute the invariant $S(\underset{\sim}{M})$, when $\underset{\sim}{M}$ is a factor.

Section 4.3 is devoted to the construction of examples of factors of all the types: $I_n$, $I_\infty$, $II_1$, $II_\infty$, $III_\lambda$ $(0 \leqslant \lambda \leqslant 1)$. Practically all these examples arise as the crossed-product of $L^\infty(X,\mathcal{F},\mu)$ by an ergodic group of automorphisms; the construction of factors of type $III_\lambda$, $\lambda \in [0,1]$, requires the construction of ergodic groups of automorphisms of a measure space, with specified "ratio sets" in the sense of Krieger.

Section 4.4 takes up the construction of the crossed-product, when $G$ is a general (not necessarily discrete) locally compact group. If $\underset{\sim}{M} = M \otimes_\alpha G$, with $G$ locally compact and abelian, an action $\tilde{\alpha}$ of $\Gamma$ on $\underset{\sim}{M}$ is constructed. The main result of this section is Takesaki's duality theorem which states that $\underset{\sim}{M} \otimes_{\tilde{\alpha}} \Gamma$ is naturally isomorphic to $M \otimes \mathcal{L}(L^2(G))$. This is a genuine duality theorem if it is the case that $M \cong M \otimes \mathcal{L}(L^2(G))$. It is shown that such is the case for a fairly large class of (the so-called properly infinite) von Neumann algebras, which includes all infinite factors.

Section 4.5 applies the results of Section 4.4 to the case when $M$ is a factor of type III, $G = \mathbb{R}$ and $\alpha = \sigma^\phi$, where $\phi$ is a fns weight on $M$.

This section contains very few proofs; instead, some heuristic arguments are given, which, it is hoped, will leave the reader with a "reasonable belief" in the result that a factor of type III is "essentially uniquely" expressible as the crossed-product of a semifinite von Neumann algebra $N$ by a one-parameter group $\{\theta_t\}$ of automorphisms of $N$ which satisfy $\tau \circ \theta_t = e^{-t}\tau$ for all $t$ in $\mathbb{R}$ for some fns weight $\tau$ on $M$.

## 4.1. Discrete Crossed-Products

Throughout this section, we shall assume that $\alpha$ is an action of a countable discrete (not necessarily abelian) group $G$ on a von Neumann algebra $M$. As at the end of Section 2.2, we shall let $\epsilon$ denote the identity element of $G$, $\{\xi_t: t \in G\}$ denote the canonical orthonormal basis for $\ell^2(G)$ (defined by $\xi_t(s) = \delta_{st}$) and $t \to \lambda_t$ (resp., $t \to \rho_t$) denote the left- (resp., right-) regular representation of $G$ in $\ell^2(G)$.

We shall assume that $M \subseteq \mathcal{L}(\mathcal{H})$ and let $\tilde{\mathcal{H}} = \oplus_{t\in G}\mathcal{H}_t$, where $\mathcal{H}_t = \mathcal{H}$ for all $t$. An element of $\tilde{\mathcal{H}}$ is a map $\tilde{\xi}: G \to \mathcal{H}$ such that $\Sigma\|\tilde{\xi}(t)\|^2 < \infty$. For $\xi$ in $\mathcal{H}$ and $t$ in $G$, we shall write $\tilde{\xi}^{(t)}(s) = \delta_{st}\xi$. There is a natural identification $\tilde{\mathcal{H}} \cong \mathcal{H} \otimes \ell^2(G)$ whereby $\tilde{\xi}^{(t)} \leftrightarrow \xi \otimes \xi_t$ for all $\xi$ in $\mathcal{H}$ and $t$ in $G$; in particular, $\{\tilde{\xi}_s^{(t)}: s,t \in G\}$ is an orthonormal basis for $\tilde{\mathcal{H}}$.

Any $\tilde{x}$ in $\mathcal{L}(\tilde{\mathcal{H}})$ is represented by a unique matrix $((\tilde{x}(s,t)))$ -- with rows and columns indexed by $G$ -- where, for each $s$ and $t$ in $G$, $\tilde{x}(s,t)$ is the unique bounded operator on $\mathcal{H}$ satisfying $\langle\tilde{x}(s,t)\xi,\eta\rangle = \langle\tilde{x}\tilde{\xi}^{(t)},\tilde{\eta}^{(s)}\rangle$ for all $\xi,\eta$ in $\mathcal{H}$.

### Exercises

**(4.1.1)**

(a) If $\tilde{x} \in \mathcal{L}(\tilde{\mathcal{H}})$ and $\tilde{\xi} \in \tilde{\mathcal{H}}$, then for $s$ in $G$,

$$(\tilde{x}\tilde{\xi})(s) = \sum_{t\in G} \tilde{x}(s,t)\tilde{\xi}(t),$$

the series on the right converging in norm in $\mathcal{H}$;

(b) If $\tilde{x} \in \mathcal{L}(\tilde{\mathcal{H}})$, then $\tilde{x}^*(s,t) = \tilde{x}(t,s)^* \quad \forall\, s,t \in G$;

(c) If $\tilde{x}, \tilde{y}, \tilde{z} \in \mathcal{L}(\tilde{\mathcal{H}})$ and $\tilde{z} = \tilde{x}\tilde{y}$, then $\forall\, s,t \in G$,

$$\tilde{z}(s,t) = \sum_{u\in G} \tilde{x}(s,u)\tilde{y}(u,t),$$

the series on the right converging $\sigma$-strongly*. (Hint: since $\|$any (finite) partial sum$\| \leqslant \max\{\|\tilde{x}\|, \|\tilde{y}\|\}$ it suffices, by (b), to prove strong convergence; for this, use (a).)

**Definition 4.1.2.** With the above notation, define

$$\widetilde{M} = \{\widetilde{x} \in \mathfrak{L}(\widetilde{\mathcal{H}}): \widetilde{x}(s,t) \in M \text{ and } \widetilde{x}(s,t) = \alpha_{t^{-1}}(\widetilde{x}(st^{-1},\epsilon)) \text{ for all } s,t \text{ in } G\}.$$

The set $\widetilde{M}$ is called the crossed-product of $M$ with $G$ (by $\alpha$) and also denoted sometimes by $M \otimes_\alpha G$. □

Note that if $\widetilde{x} \in \widetilde{M}$ and $s,t,u \in G$, then

$$\widetilde{x}(su,tu) = \alpha_{u^{-1}}(\widetilde{x}(s,t)).$$

It is easy to see that $\widetilde{M}$ is weakly closed (in $\mathfrak{L}(\widetilde{\mathcal{H}})$); it is not much harder to verify that $\widetilde{M}$ is a self-adjoint subalgebra of $\mathfrak{L}(\widetilde{\mathcal{H}})$ containing 1, and consequently a von Neumann algebra of operators on $\widetilde{\mathcal{H}}$. (For instance, if $\widetilde{x},\widetilde{y} \in \widetilde{M}$ and $\widetilde{z} = \widetilde{x}\widetilde{y}$, it follows from Ex. (4.1.1) (c) and the fact that $M$ is weakly closed, that $\widetilde{z}(s,t) \in M$; further,

$$\begin{aligned}\widetilde{z}(s,t) &= \sum_u \widetilde{x}(s,u)\widetilde{y}(u,t) = \sum_u \widetilde{x}(s,u)\alpha_{t^{-1}}(\widetilde{y}(ut^{-1},\epsilon)) \\ &= \sum_v \widetilde{x}(s,vt)\alpha_{t^{-1}}(\widetilde{y}(v,\epsilon) = \sum_v \alpha_{t^{-1}}(\widetilde{x}(st^{-1},v)\widetilde{y}(v,\epsilon)) \\ &= \alpha_{t^{-1}}(\widetilde{z}(st^{-1},\epsilon)).)\end{aligned}$$

Define $\pi: M \to \widetilde{M}$ by the prescription $(\pi(x))(s,t) = \delta_{st}\alpha_{t^{-1}}(x)$. (When there is more than one action floating around, we shall sometimes write $\pi_\alpha$ for $\pi$.) It is readily verified that $\pi$ is a normal *-algebra isomorphism of $M$ into $\widetilde{M}$. Hence $\pi(M)$ is a von Neumann subalgebra of $\widetilde{M}$; in fact it is precisely the set of those $\widetilde{x}$ in $\widetilde{M}$ for which $\widetilde{x}(s,t) = 0$ when $s \neq t$.

Next, define $\lambda: G \to \mathfrak{L}(\widetilde{\mathcal{H}})$ by letting $(\lambda(u))(s,t) = \delta_{s,ut}$, or equivalently, $(\lambda(u)\xi)(t) = \xi(u^{-1}t)$. It is clear that $\lambda$ is a unitary representation of $G$ in $\widetilde{\mathcal{H}}$; in the identification $\widetilde{\mathcal{H}} = \mathcal{H} \otimes \ell^2(G)$, $\lambda(u) = 1 \otimes \lambda_u$. It is easily verified that $\lambda(G) \subseteq \widetilde{M}$ and that

$$\text{(4.1.1)} \qquad \lambda(u)\pi(x)\lambda(u)^* = \pi(\alpha_u(x))$$

for all $u$ in $G$ and $x$ in $M$.

It follows from the above discussion that $(\pi(M) \cup \lambda(G))'' \subseteq \widetilde{M}$; the following exercises outline a proof of the reverse inclusion.

**Exercises**

**(4.1.3)**

(a) Let $\tilde{x} \in \tilde{M}$; say that $\tilde{x}$ is supported on the $u$th diagonal, for some $u$ in $G$, if $\tilde{x}(s,t) = 0$ whenever $st^{-1} \neq u$. Show that $\tilde{x}$ is supported on the $u$th diagonal if and only if $\tilde{x} = \pi(x)\lambda(u)$ for some $x$ in $M$. (When $u = \epsilon$, this has already been noted.)

(b) If $\tilde{x} \in \mathfrak{L}(\mathcal{H})$ and $u \in G$, define $\tilde{x}(u) \in \mathfrak{L}(\mathcal{H})$ by

$$(\tilde{x}(u))(s,t) = \begin{cases} \tilde{x}(s,t), & \text{if } st^{-1} = u \\ 0, & \text{if } st^{-1} \neq u. \end{cases}$$

Show that $\tilde{x} = \Sigma_{u \in G}\tilde{x}(u)$ the series on the right being interpreted as the $\sigma$-strong* limit of the net of finite sums.

(c) If $\tilde{x} \in \tilde{M}$, show that

$$\tilde{x} = \sum_{u \in G} \pi(\alpha_u(\tilde{x}(u,\epsilon)))\lambda(u),$$

the sum being interpreted as in (b); in particular $\tilde{M} = (\pi(M) \cup \lambda(G))''$; more explicitly, if $M_0$ is the set of those $\tilde{x}$ in $\tilde{M}$ which are supported on finitely many diagonals (or equivalently $M_0$ is the set of operators of the form

$$\sum_{u \in G} \pi(x(u))\lambda(u),$$

where $x: G \to M$ satisfies $x(u) = 0$ for all but finitely many $u$), then $M_0$ is a $\sigma$-strongly* dense self-adjoint subalgebra of $\tilde{M}$.

**Examples 4.1.4.** (a) Let $\mathcal{H}$ be one-dimensional and $M = \mathfrak{L}(\mathcal{H})$; thus $M \cong \mathbb{C}$. For any countable discrete group, let $\alpha_t = id_M$ for all $t$. Then $\tilde{\mathcal{H}}$ is naturally identified with $\ell^2(G)$, $\pi(\lambda 1) = \lambda 1$ and $\lambda(u)$ gets identified with $\lambda_u$. Thus, in this case $\tilde{M} = \lambda(G)''$ is just the group von Neumann algebra $W^*(G)$.

(b) If $H$ and $K$ are countable discrete groups and $\alpha : K \to \mathrm{Aut}(H)$ is a homomorphism, recall that the semidirect product, which we shall denote by $H \otimes_\alpha K$, is the group $G$ whose underlying set is $H \times K$, where group-multiplication is given by

$$(h_1,k_1)(h_2,k_2) = (h_1\alpha_{k_1}(h_2),k_1k_2);$$

an easy computation reveals that

$$(*) \qquad (h_0,k_0)^{-1}(h,k) = (\alpha_{k_0^{-1}}(h_0^{-1}h),\ k_0^{-1}k).$$

It is easily verified that the equation $(u_k\xi)(h) = \xi(\alpha_{k^{-1}}(h))$, $h \in H$,

$\xi \in \ell^2(H)$ defines a unitary representation $k \to u_k$ of $K$ in $\ell^2(H)$; if $\lambda_H$ denotes the left-regular representation of $H$, it is easy to check that $u_k\lambda_H(h)u_k^* = \lambda_H(\alpha_k(h))$ for $k$ in $K$ and $h$ in $H$. It follows that there exists an action $\tilde{\alpha}$ of $K$ on $W^*(H)$ such that $\tilde{\alpha}_k(x) = u_k x u_k^*$ for all $x$ in $W^*(H)$ and $k$ in $K$.

The crossed-product $W^*(H) \otimes_{\tilde{\alpha}} K$ acts on the Hilbert space $\widetilde{\mathcal{H}} = \ell^2(K; \ell^2(H))$ which can be naturally identified with $\ell^2(H \times K) = \ell^2(G)$. Under this identification,

$$(\pi_{\tilde{\alpha}}(\lambda_H(h_0))\tilde{\xi})(h,k) = \tilde{\xi}(\alpha_{k^{-1}}(h_0^{-1})h,k)$$

and

$$(\lambda(k_0)\tilde{\xi})(h,k) = \tilde{\xi}(h, k_0^{-1}k)$$

for $h_0 \in H$, $k_0 \in K$ and $\tilde{\xi}$ in $\ell^2(G)$.

If $w$ denotes the (clearly unitary) operator on $\ell^2(G)$ defined by

$$(w\tilde{\xi})(h,k) = \tilde{\xi}(\alpha_{k^{-1}}(h),k),$$

it follows that

$$(w\pi_{\tilde{\alpha}}(\lambda_H(h_0))w^*\tilde{\xi})(h,k) = \tilde{\xi}(h_0^{-1}h, k)$$

and

$$(w\lambda(k_0)w^*\tilde{\xi})(h,k) = \tilde{\xi}(\alpha_{k_0^{-1}}(h), k_0^{-1}k);$$

in view of equation (*), this says that

$$w\pi_{\tilde{\alpha}}(\lambda_H(h_0))w^* = \lambda_G(h_0,\epsilon_K)$$

and

$$w\lambda(k_0)w^* = \lambda_G(\epsilon_H, k_0);$$

thus

$$w(W^*(H) \otimes_{\tilde{\alpha}} K)w^* = W^*(H \otimes_\alpha K). \qquad \square$$

**Remark 4.1.5.** Although the construction of the crossed-product seems to depend upon the Hilbert space $\mathcal{H}$ on which $M$ acts, it is a fact -- which we shall prove in Section 4.4, when dealing with continuous crossed-products -- that the isomorphism class of $M$ depends only on the isomorphism class of the dynamical system $(M,G,\alpha)$; explicitly, if $(M_i,G,\alpha_i)$, $i = 1,2$, are dynamical systems and if $\pi$: $M_1 \to M_2$ is a von Neumann algebra isomorphism such that $\pi \circ \alpha_{1,t} = \alpha_{2,t} \circ \pi$ for all $t$, then

$$M_1 \otimes_{\alpha_1} G \cong M_2 \otimes_{\alpha_2} G. \qquad \square$$

**Exercises**

**(4.1.6)** Define $E: \widetilde{M} \to \pi(M)$ by $E\tilde{x} = \pi(\tilde{x}(\epsilon,\epsilon))$.

(a) Prove that $E$ is a faithful, normal norm-one projection of $\widetilde{M}$ onto the von Neumann subalgebra $\pi(M)$.

(b) If $\phi$ is any fns weight on $M$, define $\tilde{\phi}(\tilde{x}) = \phi(\tilde{x}(\epsilon,\epsilon))$ for $\tilde{x}$ in $\widetilde{M}_+$; show that $\tilde{\phi}$ is a fns weight on $\widetilde{M}$. (Hint: for semifiniteness, use the semifiniteness of $\phi$ via Ex. (2.4.8) (d).) $\square$

We turn next to a discussion of conditions on the dynamical system $(M,G,\alpha)$ which ensure that $\widetilde{M}$ is a factor.

**Lemma 4.1.7.** *Let* $\tilde{x} \in \widetilde{M}$.

(a) $\tilde{x} \in \pi_\alpha(M)' \Leftrightarrow \tilde{x}(t,\epsilon)y = \alpha_{t^{-1}}(y)\tilde{x}(t,\epsilon)$ *for all* $y$ *in* $M$ *and* $t$ *in* $G$;

(b) $\tilde{x} \in \lambda(G)' \Leftrightarrow \tilde{x}(utu^{-1},\epsilon) = \alpha_u(\tilde{x}(t,\epsilon))$ *for all* $t$, $u$ *in* $G$.

**Proof.** (a) Let $y \in M$. Then

$$\tilde{x}\pi(y) = \pi(y)\tilde{x} \Leftrightarrow (\tilde{x}\pi(y))(t,\epsilon) = (\pi(y)\tilde{x})t,\epsilon) \quad \forall t \text{ in } G$$

$$\Leftrightarrow \tilde{x}(t,\ \epsilon)y = \alpha_{t^{-1}}(y)\tilde{x}(t,\ \epsilon) \quad \forall t \text{ in } G,$$

and the assertion follows.

(b) Let $u \in G$. Then

$$\tilde{x}\lambda(u) = \lambda(u)\tilde{x}$$

$$\Leftrightarrow (\tilde{x}\lambda(u))(s,\ \epsilon) = (\lambda(u)\tilde{x})(s,\ \epsilon) \quad \forall s \text{ in } G$$

$$\Leftrightarrow \tilde{x}(s,u) = \tilde{x}(u^{-1}s,\ \epsilon) \quad \forall s \text{ in } G$$

$$\Leftrightarrow \alpha_{u^{-1}}(\tilde{x}(su^{-1},\ \epsilon)) = \tilde{x}(u^{-1}s,\ \epsilon) \quad \forall s \text{ in } G$$

$$\Leftrightarrow \alpha_{u^{-1}}(\tilde{x}(utu^{-1},\ \epsilon)) = \tilde{x}(t,\ \epsilon) \quad \forall t \text{ in } G$$

(on putting $t = u^{-1}s$), and the proof is complete. $\square$

**Definition 4.1.8.** (a) An automorphism $\theta$ of $M$ is said to be free if $xy = \theta(y)x$ for all $y$ in $M$ implies $x = 0$.

(b) An action $\alpha$ of $G$ on $M$ is said to be free if for $t \neq \epsilon$, the automorphism $\alpha_t$ is free, as in (a) above. $\square$

**Corollary 4.1.9.** *The action* $\alpha$ *is free if and only if*

$$\widetilde{M} \cap \pi_\alpha(M)' = \pi_\alpha(Z(M)).$$

**Proof.** Suppose the action $\alpha$ is free. Then, it follows from Lemma 4.1.7 (a) that

$$\widetilde{x} \in \widetilde{M} \cap \pi_\alpha(M)' \Rightarrow \widetilde{x}(t,\epsilon) = 0 \quad \text{for } t \neq \epsilon$$

$$\Rightarrow \widetilde{x} = \pi(\widetilde{x}(\epsilon,\epsilon)) \in \pi(M).$$

Since $\pi_\alpha$ is 1-1, the assumption $\widetilde{x} \in \pi_\alpha(M)'$ forces $\widetilde{x}(\epsilon,\epsilon)$ to belong to $Z(M)$, and so, $\widetilde{M} \cap \pi_\alpha(M)' \subseteq \pi_\alpha(Z(M))$; the other inclusion is obvious.

If conversely, there exists $t \neq \epsilon$ such that $\alpha_t$ is not free, then (by definition) there exists a non-zero $x$ in $M$ such that $xy = \alpha_t(y)x$ for all $y$ in $M$; define $\widetilde{x}$ in $\widetilde{M}$ by $\widetilde{x}(s^{-1},\epsilon) = \delta_{st}x$ and notice, that, by Lemma 4.1.7 (a), $\widetilde{x} \in \widetilde{M} \cap \pi_\alpha(M)'$ while $\widetilde{x} \notin \pi_\alpha(M)$. □

The use of the adjective "free" in Definition 4.1.8 stems from the case when $M = L^\infty(X,\mathcal{F},\mu)$. Before seeing this, let us digress with another definition.

**Definition 4.1.10.** Let $(X,\mathcal{F},\mu)$ be a separable $\sigma$-finite measure space. By an automorphism of $(X,\mathcal{F},\mu)$ is meant a bimeasurable bijective $\mu$-null-sets preserving transformation of $X$; i.e., an automorphism is a bijection $T$: $X \to X$ such that (i) $E \in \mathcal{F}$ implies $T(E)$, $T^{-1}(E) \in \mathcal{F}$; and (ii) if $E \in \mathcal{F}$, then $\mu(E) = 0$ if and only if $\mu(T^{-1}(E)) = 0$. □

**Example 4.1.11.** Each automorphism $T$ of $(X,\mathcal{F},\mu)$ induces an automorphism $\alpha_T$ of $M = L^\infty(X,\mathcal{F},\mu)$ thus: $\alpha_T(f) = f \circ T^{-1}$.

**Assertion:** $\alpha_T$ is free $\Leftrightarrow$ whenever $E \in \mathcal{F}$ and $\mu(E) > 0$, $\exists F \in \mathcal{F}$ such that $F \subseteq E$, $\mu(F) > 0$ and $F \cap TF = \phi$.

**Proof of Assertion.** $\Leftarrow$ : If $fg = \alpha_T(g)f$ for all $g$ in $M$, it follows that if $E = \{\omega \in X: f(\omega) \neq 0\}$, then $1_E g = 1_E \cdot (g \circ T^{-1})$ for all $g$ in $L^\infty(X,\mathcal{F},\mu)$ -- as usual, equality means $\mu$ - a.e. equality. If $f \neq 0$ and $\mu(E) > 0$, and if $F$ is as in the assertion, the above equality is violated if $g = 1_F$; this contradiction establishes that $\alpha_T$ is free.

$\Rightarrow$ : If we can find $F_0$ in $\mathcal{F}$ such that $F_0 \subseteq E$ and $\mu(F_0 \backslash T^{-1}(F_0)) > 0$, the set $F = F_0 \backslash T^{-1}(F_0)$ does the needful; if no such set $F_0$ exists, argue that it must be the case that $\mu((T(F) \cap E)\Delta F) = 0$ whenever $F \in \mathcal{F}$, $F \subseteq E$ and $\mu(F) > 0$; hence $1_E f = 1_E \cdot (f \circ T^{-1})$ a.e., whenever $f = 1_F$ with $F$ as above; conclude that the above equation persists for all $f$ in $M$, thereby contradicting the assumption that $\alpha_T$ is free. □

**Exercises**

**(4.1.12)** If the Borel space $(X,\mathcal{F})$ is countable separated -- i.e., if there exists a sequence $\{E_n\}$ in $\mathcal{F}$ such that for any $\omega_1,\omega_2 \in X$,

$$\omega_1 = \omega_2 \Leftrightarrow 1_{E_n}(\omega_1) = 1_{E_n}(\omega_2) \quad \text{for all } n$$

-- show that $\alpha_T$ is free (with the notation of Example 4.1.11) $\Leftrightarrow$ $\mu(\{\omega \in X: T\omega = \omega\}) = 0$. □

**Definition 4.1.13.** An action $\alpha$ of $G$ on $M$ is said to be ergodic if $M^{\alpha} = \{\lambda\cdot 1: \lambda \in \mathbb{C}\}$.

**Example 4.1.14.** Suppose $t \to T_t$ is a homomorphism from $G$ into the group of automorphisms of $(X,\mathcal{F},\mu)$; we then have an induced action $\alpha$ of $G$ on $M = L^{\infty}(X,\mathcal{F},\mu)$ given by $\alpha_t(f) = f \circ T_t^{-1}$. (It is a fact -- which we will not pursue further -- that if $(X,\mathcal{F},\mu)$ is separable and $\sigma$-finite, then every action of $G$ on $M$ is obtained in this fashion.) It is, then, not hard to verify (do it!) that the action $\alpha$ is ergodic $\Leftrightarrow$ $E \in \mathcal{F}$, $\mu(T_t(E) \,\Delta\, E) = 0$ for all $t$ in $G$ implies $\mu(E) = 0$ or $\mu(X\backslash E) = 0$ -- that is the classical notion of ergodicity. □

**Proposition 4.1.15.** *Suppose $\alpha$ is a free action of $G$ on $M$. Then $\alpha_t(Z(M)) = Z(M)$ for all $t$; let $\alpha_Z$ denote the induced action (by restriction) of $G$ on $Z(M)$. Then, the following conditions are equivalent:*

(i) *$M \otimes_{\alpha} G$ is a factor;*
(ii) *$\alpha_Z$ is an ergodic action of $G$ on $Z(M)$.*

**Proof.** The first assertion is clear since any automorphism of $M$ will map $Z(M)$ onto itself. Let $\tilde{x} \in Z(\tilde{M})$. It follows from Corollary 4.1.9 that $\tilde{x} = \pi_{\alpha}(x)$ for some $x$ in $Z(M)$. Notice now that for any $u$ in $G$,

$$\begin{aligned} \pi_{\alpha}(x) &= \tilde{x} = \lambda(u)\tilde{x}\lambda(u)^* \\ &= \lambda(u)\pi_{\alpha}(x)\lambda(u^*) \\ &= \pi_{\alpha}(\alpha_u(x)) \qquad \text{(cf. eqn. (4.1.1))} \end{aligned}$$

and hence $x = \alpha_u(x)$ for all $u$ in $G$. Thus, we have shown that

$$Z(\tilde{M}) \subseteq \{\pi_{\alpha}(x): x \in Z(M),\ \alpha_u(x) = x \ \ \forall u \text{ in } G\};$$

since the other inclusion is clear, it follows that $Z(\tilde{M}) = \pi_{\alpha}(Z(M)^{\alpha_Z})$; the equivalence of (i) and (ii) follows immediately. □

For convenience of reference, and for want of a better place to locate it, we state the following result here.

**Proposition 4.1.16.** *Let $\theta$ be an automorphism of a factor $M$; the following conditions on $\theta$ are equivalent:*

(i) $\theta$ *is free;*
(ii) $\theta$ *is outer -- i.e., there does not exist a $u$ in $\mathcal{U}(M)$ such that $\theta(x) = uxu^*$ for all $x$ in $M$.*

**Proof.** (i) $\Rightarrow$ (ii): If $\theta$ is inner, i.e., if there exists $u$ in $\mathcal{U}(M)$ such that $\theta(y) = uyu^*$ for all $y$ in $M$, then $uy = \theta(y)u$ for all $y$ in $M$, whence $\theta$ is not free.

(ii) $\Rightarrow$ (i): Suppose $x \in M$ and $x$ satisfies $xy = \theta(y)x$ for all $y$ in $M$; then $y^*x^* = x^*\theta(y^*)$ for all $y$ in $M$; so, if $v \in \mathcal{U}(M)$, note that $v^*x^*xv = x^*\theta(v^*v)x = x^*x$, and $\theta(v)xx^*\theta(v)^* = xvv^*x^* = xx^*$; since $v \in \mathcal{U}(M)$ was arbitrary, conclude (via Scholium 0.4.8) that $x^*x, xx^* \in Z(M)$. Since $M$ is a factor, it follows that $x = \|x\|u$, for some $u$ in $\mathcal{U}(M)$; if $x \neq 0$, infer that for any $y$ in $M$, $uy = \theta(y)u$, or $\theta(y) = uyu^*$, contradicting the assumption that $\theta$ is not inner. □

## 4.2. The Modular Operator for a Discrete Crossed-Product

Throughout this section, we shall let $(M,G,\alpha)$ be a discrete dynamical system, $\tilde{M}$ denote the crossed-product $M \otimes_\alpha G$, and let $\tilde{\phi}$ denote the fns weight on $\tilde{M}$ (cf. Ex. (4.1.6) (b)) induced by a fns weight $\phi$ on $M$ by the equation $\tilde{\phi}(\tilde{x}) = \phi(\tilde{x}(\epsilon,\epsilon))$. In this section, we shall attempt to determine the GNS space and the "modular operators $S$, $F$, $J$, $\Delta$" associated with $(\tilde{M},\tilde{\phi})$ in terms of the corresponding objects for $(M,\phi)$. In view of Remark 4.1.5, we shall assume that $M$ is "standard with respect to $\phi$" -- i.e., that $M \subseteq \mathcal{L}(\mathcal{H})$, $\mathcal{H} = \mathcal{H}_\phi$ and $\pi_\phi(x) = x$ for all $x$ in $M$. We shall see that in this case $\tilde{M}$ is "standard with respect to $\tilde{\phi}$", by explicitly constructing $\eta_{\tilde{\phi}}$; for notational convenience, we shall write $\eta$ and $\tilde{\eta}$ in place of $\eta_\phi$ and $\eta_{\tilde{\phi}}$, respectively; likewise, we shall write $\mathfrak{D}$, $\mathfrak{N}$, $\mathfrak{M}$ and $\tilde{\mathfrak{D}}$, $\tilde{\mathfrak{N}}$, $\tilde{\mathfrak{M}}$ instead of $\mathfrak{D}_\phi$, $\mathfrak{N}_\phi$, $\mathfrak{M}_\phi$ and $\mathfrak{D}_{\tilde{\phi}}$, $\mathfrak{N}_{\tilde{\phi}}$, $\mathfrak{M}_{\tilde{\phi}}$ respectively.

We shall begin the proceedings by a technical lemma which states that the linear map $\eta\colon \mathfrak{N} \to \mathcal{H}$ is closed in a certain sense.

**Lemma 4.2.1.** *If $\{x_i\}$ is a net in $\mathfrak{N}$ such that $\sup_i \|x_i\| < \infty$, $x_i \to x$ $\sigma$-strongly* and $\eta(x_i) \to \xi$ in $\mathcal{H}$, then $x \in \mathfrak{N}$ and $\xi = \eta(x)$.*

**Proof.** Pass to a subnet (in fact of the form $\{x_i: i \geqslant i_0\}$) and assume that $\|\eta(x_i)\| < \|\xi\| + 1$ for all $i$. Then there exists a constant $C > 0$ such that $\|x_i\| \leqslant C$ and $\|\eta(x_i)\| \leqslant C$ for all $i$; it follows as in the proof of Lemma 3.4.4, that the set $K = \{y \in \mathfrak{N}: \|y\| \leqslant C, \|\eta(y)\| \leqslant C\}$ is $\sigma$-strongly* closed and consequently $x \in \mathfrak{N}$.

For any $y$ in $\mathfrak{N}^*$, notice that

$$\eta(yx_i) = y\eta(x_i) \to y\xi ,$$

$$S_0\eta(yx_i) = x_i^*\eta(y^*) \to x^*\eta(y^*).$$

Since $S$ is closed, conclude that $y\xi \in \operatorname{dom} S$ and $S(y\xi) = x^*\eta(y^*) = \eta(x^*y^*) = S\eta(yx) = Sy\eta(x)$; the injectivity of $S$ forces $y\xi = y\eta(x)$; since $y \in N^*$ was arbitrary, conclude from the semifiniteness of $\phi$ that $\xi = \eta(x)$. □

**Proposition 4.2.2.** *Retain the above notation.*

(a) $\tilde{x} \in \tilde{N} \Leftrightarrow \tilde{x}(s,\epsilon) \in N$ *for all s in G, and*

$$\sum_{s\in G} \|\eta(\tilde{x}(s,\epsilon))\|^2 < \infty ;$$

(b) *define* $\tilde{\eta}$: $\tilde{N} \to \tilde{\mathcal{H}}$ $(= \ell^2(G;\mathcal{H}))$ by $(\tilde{\eta}(\tilde{x}))(s) = \eta(\tilde{x}(s,\epsilon))$; then $(\tilde{\mathcal{H}}, id_{\tilde{M}}, \tilde{\eta})$ *is a* GNS *triple for* $(\tilde{M},\tilde{\phi})$.

**Proof.** (a) If $\tilde{x} \in \tilde{M}$, then

$$(\tilde{x}^*\tilde{x})(\epsilon,\epsilon) = \sum_{s\in G} \tilde{x}(s,\epsilon)^*\tilde{x}(s,\epsilon).$$

the series being "unconditionally $\sigma$-strongly* convergent" (cf. Ex. (4.1.1)). Since $\phi$ is normal, conclude that

$$\tilde{\phi}(\tilde{x}^*\tilde{x}) = \sum_{s\in G} \phi(\tilde{x}(s,\epsilon)^*\tilde{x}(s,\epsilon));$$

the assertion (a) follows, and we further have

$$(*) \qquad \tilde{\phi}(\tilde{x}^*\tilde{x}) = \sum_{s\in G} \|\eta(\tilde{x}(s, \epsilon)\|^2, \quad \tilde{x} \in \tilde{N} .$$

(b) If $\tilde{x} \in \tilde{N}$, the equation (*) ensures that the equation $(\tilde{\eta}(\tilde{x}))(s) = \eta(\tilde{x}(s,\epsilon))$ does define a map $\tilde{\eta}$: $\tilde{N} \to \tilde{\mathcal{H}}$, which is clearly linear; by definition of $\tilde{\eta}$, and equation (*), we have $\|\tilde{\eta}(\tilde{x})\|^2 = \tilde{\phi}(\tilde{x}^*\tilde{x})$; it follows, by polarization, that for any $\tilde{x}$, $\tilde{y}$ in $\tilde{N}$,

$$\langle\tilde{\eta}(\tilde{x}), \tilde{\eta}(\tilde{y})\rangle = \tilde{\phi}(\tilde{y}^*\tilde{x}).$$

The assumed density of $\eta(N)$ in $\mathcal{H}$, and an easy approximation argument shows that $\tilde{\eta}(\tilde{N})$ -- in fact $\tilde{\eta}(\tilde{N} \cap M_0)$, with $M_0$ as in Ex. (4.1.3) (c) -- is dense in $\tilde{\mathcal{H}}$.

To complete the proof, we must show that if $\tilde{x} \in \tilde{M}$ and $\tilde{y} \in \tilde{N}$, then $\tilde{x}\tilde{\eta}(\tilde{y}) = \tilde{\eta}(\tilde{x}\ \tilde{y})$. To do this, fix an $s$ in $G$, and let $\{t_1,t_2, ...\}$ be an enumeration of $G$. For $N = 1,2, ...,$ define $\tilde{x}_N$ in $\mathcal{L}(\tilde{\mathcal{H}})$ by the equation

$$\tilde{x}_N(s,t) = \tilde{x}(s,t)1_{\{t_1,...,t_N\}}(t);$$

then $\tilde{x}_N = \tilde{x}\ \tilde{p}_N$, where $\tilde{p}_N$ is the projection defined by

$$\tilde{p}_N(s,t) = \delta_{st} 1_{\{t_1,\ldots,t_N\}}(t);$$

in particular, $\|\tilde{x}_N\| \leqslant \|\tilde{x}\|$. Let

$$z_N = \sum_{n=1}^{N} \tilde{x}(s,t_n)\tilde{y}(t_n,\epsilon)$$

and observe that:

(i) $$\|z_N\| = \|\sum_{t\in G} \tilde{x}_N(s,t)\tilde{y}(t,\epsilon)\|$$
$$= \|(\tilde{x}_N\tilde{y})(s,\epsilon)\|$$
$$\leqslant \|\tilde{x}_N\tilde{y}\|$$
$$\leqslant \|\tilde{x}\| \; \|\tilde{y}\| \; ;$$

(ii) $z_N \to (\tilde{x}\ \tilde{y})(s,\epsilon)$ $\sigma$-strongly* (by Ex. (4.1.1) (c));

(iii) $z_N \in N$ and $$\eta(z_N) = \sum_{n=1}^{N} \tilde{x}(s,t_n)\eta(\tilde{y}(t_n,\epsilon))$$
$$= \sum_{n=1}^{N} \tilde{x}(s,t_n)(\tilde{\eta}(\tilde{y}))(t_n)$$
$$\to (\tilde{x}\ \tilde{\eta}(\tilde{y}))(s) \quad \text{(by Ex. (4.1.1) (a)).}$$

Conclude from Lemma 4.2.1 that

$$(\tilde{x}\ \tilde{\eta}(\tilde{y}))(s) = \eta((\tilde{x}\ \tilde{y})(s,\epsilon)) = (\tilde{\eta}(\tilde{x}\ \tilde{y}))(s);$$

since $s$ was arbitrary, the proof is complete. □

Assume for the rest of this section that $M$ is semifinite and that $\phi$ is a fns trace on $M$. (We have chosen not to denote the trace by $\tau$ for the following reasons: (a) if $\tilde{\tau}$ is the induced weight on $\tilde{M}$, then $\tilde{\tau}$ need not be a trace, and this could be confusing; (b) we can continue to use the notation established so far in this section.) Recall that $\eta$ always means $\eta_\phi$ and $\tilde{\eta}$ means $\eta_{\tilde{\phi}}$.

**Lemma 4.2.3.** *Let $\tau$ be another* fns *trace on $M$.*

(a) *There exists a unique invertible positive self-adjoint operator $H \eta M$ such that $\tau = \phi(H,\cdot)$, as in Theorem 2.6.3;*
(b) $H \ \eta \ Z(M)$;
(c) *$\eta(N_\phi \cap N_\tau)$ is a core for $H^{1/2}$;*
(d) *if $x,y \in N_\phi \cap N_\tau$, then*

$$\tau(y^*x) = \langle H^{1/2}\eta(x), H^{1/2}\eta(y)\rangle \, .$$

**Proof.** (a) Since $\sigma_t^\phi = id_M$ for all $t$, the assertion follows immediately from Theorem 2.6.3, the invertibility of $H$ being a consequence of the faithfulness of $\tau$.

(b) We know from Theorem 3.1.10 that $\sigma_t^\tau(x) = H^{it}xH^{-it}$; since $\tau$ is a trace, conclude that $H^{it} \in Z(M)$ for all $t$, whence $H \, \eta \, Z(M)$.

(c) Let $e_n = 1_{[0,n]}(H)$ $(\in Z(M))$, and define

$$\mathfrak{D}_0 = \bigcup_{n=1}^{\infty} e_n \, N .$$

Since $N$ is a left-ideal in $M$ -- actually the traciality of $\phi$ implies $N = N^*$, so that $N$ is a two-sided ideal in $M$ -- it is clear that $\mathfrak{D}_0 \subseteq N$. Recall that, for $\epsilon > 0$, $H_\epsilon = H(1 + \epsilon H)^{-1}$, note that $H_\epsilon \in Z(M)$ for all $\epsilon > 0$, and that $\tau(x) = \lim_{\epsilon \downarrow 0} \phi(H_\epsilon x)$ for $x$ in $M_+$. If $x \in N$ and $n = 1,2, \ldots,$

$$\begin{aligned} \tau((e_n x)^*(e_n x)) &= \lim_{\epsilon \downarrow 0} \phi(H_\epsilon \, x^* \, e_n \, x) \\ (\dagger) \qquad &= \lim_{\epsilon \downarrow 0} \phi(x^* \, H_\epsilon \, e_n \, x) \\ &= \langle He_n \, \eta(x), \eta(x) \rangle \\ &< \infty , \end{aligned}$$

since $H_\epsilon e_n \nearrow He_n$ as $\epsilon \downarrow 0$, and $He_n$ is bounded. Thus, $D_0 \subseteq N_\tau$, so $\mathfrak{D}_0 \subseteq N \cap N_\tau$. We shall now prove a statement that clearly implies (c): $\eta(\mathfrak{D}_0)$ is a core for $f(H)$, for any continuous function $f$ on $[0,\infty)$. (Note: the argument for (d) shows that $\eta(N \cap N_\tau) \subseteq$ dom $H^{1/2}$.)

Since $\cup_{n=1}^{\infty} e_n \mathcal{H}$ is a core for $f(H)$, it suffices to show that if $\xi \in e_N \mathcal{H}$ for some $N$, then there exists $\{\xi_n\} \subseteq \eta(\mathfrak{D}_0)$ such that $\xi_n \to \xi$ and $f(H)\xi_n \to f(H)\xi$; for this, pick $x_n$ in $N$ such that $\eta(x_n) \to \xi$, notice that $y_n = e_N x_n \in \mathfrak{D}_0$, $\eta(y_n) \in$ dom $f(H)$ and $\eta(y_n) \to \xi$; since $f(H)|e_N\mathcal{H}$ is bounded, also $f(H)\eta(y_n) \to f(H)\xi$, as desired.

(d) It suffices, thanks to polarization, to prove (d) when $x = y$. First, if $x \in \mathfrak{D}_0$ (as in (c)), so that $x = e_n x$ for some $n$, we have, by $(\dagger)$,

$$\begin{aligned} \tau(x^* x) &= \langle He_n \, \eta(x), \eta(x) \rangle \\ &= \langle (He_n)^{1/2} \, \eta(x), (He_n)^{1/2} \, \eta(x) \rangle \\ &= \langle H^{1/2} \, \eta(x), H^{1/2} \, \eta(x) \rangle, \qquad \text{as desired.} \end{aligned}$$

For a general $x$ in $N \cap N_\tau$, we have, for each $n$, by the above equation applied to $e_n x = x e_n$,

$$\tau(e_n x^* x e_n) = \langle H^{1/2} \, e_n \, \eta(x), H^{1/2} \, e_n \, \eta(x) \rangle;$$

since $e_n x^* x e_n = x^* e_n x \nearrow x^* x$, the normality of $\tau$ ensures that

$$\sup_n \|H^{1/2}e_n\ \eta(x)\|^2 = \tau(x^*x) < \infty, \quad \text{since } x \in N_\tau;$$

it follows that $\eta(x) \in \operatorname{dom} H^{1/2}$ and that $\|H^{1/2}\eta(x)\|^2 = \tau(x^*x)$. □

Coming back to the dynamical system $(M,G,\alpha)$, it is clear that the property of being a fns trace is inherited from $\phi$ by each $\phi \circ \alpha_t$; for each $t$ in $G$, let $H_t$ be the positive invertible self-adjoint operator affiliated to $Z(M)$ (as in Lemma 4.2.3) such that $\phi \circ \alpha_t = \phi(H_t\cdot)$. Let $\tilde{\phi}$, $\tilde{N}$, $\tilde{M}$, $\tilde{\mathcal{H}}$, $\eta$ be as in Proposition 4.2.2.

**Lemma 4.2.4.** *Let $\tilde{x} \in \tilde{M}$. Then*

(a) $\tilde{x} \in \hat{N} \Leftrightarrow \tilde{x}(t,\epsilon) \in N \quad \forall t$, *and* $\sum_t \|\eta(\tilde{x}(t,\epsilon))\|^2 < \infty$;

(b) $\tilde{x} \in \widehat{N^*} \Leftrightarrow \tilde{x}(t,\epsilon) \in N_{\phi \circ \alpha_t} \ \forall t$, *and*

$$\sum_t \|\eta_{\phi\circ\alpha_t}(\tilde{x}(t,\epsilon))\|^2 < \infty;$$

(c) $\tilde{x} \in \hat{N} \cap \widehat{N^*} \Leftrightarrow \tilde{x}(t,\epsilon) \in N \cap N_{\phi\circ\alpha_t} \ \forall t$ *in* $G$, *and*

$$\sum_t (\|\eta(\tilde{x}(t,\epsilon))\|^2 + \|H_t^{1/2}\eta(\tilde{x}(t,\epsilon))\|^2) < \infty.$$

**Proof.** (a) This is just Prop. 4.2.2 (a).

(b) Since $\tilde{x}^*(t,\epsilon) = \tilde{x}(\epsilon,t)^* = \alpha_{t^{-1}}(\tilde{x}(t^{-1},\epsilon)^*)$, conclude from (a) that

$$\begin{aligned}\tilde{x} \in \widehat{N^*} &\Leftrightarrow \alpha_{t^{-1}}(\tilde{x}(t^{-1},\epsilon)^*) \in N \quad \text{for all } t \\ &\qquad \text{and } \sum_t \|\eta(\alpha_{t^{-1}}(\tilde{x}(t^{-1},\epsilon))^*)\|^2 < \infty \\ &\Leftrightarrow \alpha_t(\tilde{x}(t,\epsilon)) \in N \quad \text{for all } t \\ &\qquad \text{and } \sum_t \|\eta(\alpha_t(\tilde{x}(t,\epsilon)))\|^2 < \infty\end{aligned}$$

(since $\phi$ a trace $\Rightarrow N = N^*$ and $\|\eta(y)\| = \|\eta(y^*)\|$), which is just a reformulation of (b).

(c) This is an immediate consequence of (a) and (b) of this lemma, and Lemma 4.2.3 (c) (applied to $\tau = \phi \circ \alpha_t$, for each $t$ in $G$). □

**Proposition 4.2.5.** *With the notation established thus far in this section, the modular operator $\Delta_{\tilde{\phi}}$ is given by $\Delta_{\tilde{\phi}} = \oplus_{t\in G} H_t$; i.e.,*

$$\operatorname{dom} \Delta_{\tilde{\phi}} = \left\{\tilde{\xi} \in \tilde{\mathcal{H}}: \tilde{\xi}(t) \in \operatorname{dom} H_t \ \forall t \text{ and } \sum_t \|H_t\tilde{\xi}(t)\|^2 < \infty\right\}$$

*and* $(\Delta_{\tilde{\phi}}\tilde{\xi})(t) = H_t\tilde{\xi}(t)$ *for* $\tilde{\xi}$ *in* $\operatorname{dom} \Delta_{\tilde{\phi}}$.

**Proof.** For each $t$ in $G$, it is clear that the equation $\tau_t = \phi(H_t^2\cdot)$ defines a fns trace on $M$. (Faithfulness follows from the invertibility of $H_t^2$ and the rest follows from $H_t^2 \eta Z(M)$.) Define

$$\mathfrak{D}_0 = \{\tilde{\eta}(\tilde{x}): \tilde{x} \in \check{N} \cap \check{M}_0, \ \tilde{x}(t, \epsilon) \in N_{T_t} \ \forall t \text{ in } G\},$$

with $\check{M}_0$ as in Ex. (4.1.3) (c); alternatively,

$$\mathfrak{D}_0 = \{\check{\xi} \in \check{\mathfrak{H}}: \check{\xi}(t) \in \eta(N \cap N_{T_t}) \ \forall t \text{ and } \check{\xi}(t) = 0$$
$$\text{for all but finitely many } t\}.$$

If $\check{H} = \oplus H_t$ as in Ex. 2.5.6, it follows from Lemma 4.2.3 (c), Ex. 2.5.6 (b) and the second description of $\mathfrak{D}_0$ given above, that $\mathfrak{D}_0$ is a core for $H$; it suffices, by Ex. 2.5.5 (c), to show now that $H|\mathfrak{D}_0 \subseteq \Delta_{\check{\phi}}$.

So, let $\xi \in \mathfrak{D}_0$; thus $\xi = \eta(\tilde{x})$, where $\tilde{x} \in M$, $\tilde{x}(t, \epsilon) \in N \cap N_{T_t}$ for all $t$ in $G$, and $\tilde{x}(t,\epsilon) \neq 0$ for at most finitely many $t$ in $G$. For each $t$, by Lemma 4.2.3 (c), $\xi(t) \in \text{dom } H_t \subseteq \text{dom } H_t^{1/2}$; so $\eta(\tilde{x}(t,\epsilon)) \in \text{dom } H_t^{1/2}$ and consequently

$$\tilde{x}(t, \epsilon) \in N_{\phi(H_t \cdot)} = N_{\phi o \alpha_t};$$

since $\tilde{x} \in \check{M}_0$, it follows from Lemma 4.2.4 (c) that $\tilde{x} \in \check{N} \cap \check{N}^*$, so that $\xi \in \text{dom } S$. To complete the proof it will be sufficient to establish that $\eta(\tilde{x}^*) \in \text{dom } F$ and that $F\eta(\tilde{x}^*) = \check{H}\xi$; since $\eta(N \cap \check{N}^*)$ is a core for $S$, we have, thus, to prove that if $\tilde{y} \in N \cap N^*$, then

$$<\tilde{\eta}(\tilde{y}^*), \tilde{\eta}(\tilde{x}^*)> = <\check{H} \tilde{\eta}(\tilde{x}), \tilde{\eta}(\tilde{y})>.$$

So, take such a $\tilde{y}$ and compute as follows:

$$\begin{aligned}
<\tilde{\eta}(\tilde{y}^*), \tilde{\eta}(\tilde{x}^*)> &= \sum_{t \in G} <\eta(\tilde{y}^*(t, \epsilon)), \eta(\tilde{x}^*(t, \epsilon))> \\
&= \sum_{t \in G} <\eta(\alpha_{t^{-1}}(\tilde{y}(t^{-1}, \epsilon)^*)), \eta(\alpha_{t^{-1}}(\tilde{x}(t^{-1},\epsilon)^*))> \\
&= \Sigma\, \phi \circ \alpha_{t^{-1}}(\tilde{x}(t^{-1}, \epsilon)\tilde{y}(t^{-1}, \epsilon)^*) \\
&= \Sigma\, \phi \circ \alpha_t(\tilde{x}(t, \epsilon)\tilde{y}(t, \epsilon)^*) \\
&= \Sigma\, \phi \circ \alpha_t(\tilde{y}(t, \epsilon)^*\tilde{x}(t, \epsilon)) \text{ (since } \phi \text{ is a trace)} \\
&= \Sigma <H_t^{1/2}\eta(\tilde{x}(t, \epsilon)), H_t^{1/2}\eta(\tilde{y}(t, \epsilon))> \\
&\qquad \text{(by Lemma 4.2.3 (d))} \\
&= \Sigma <H_t\eta(\tilde{x}(t, \epsilon)), \eta(\tilde{y}(t, \epsilon))> \\
&\qquad \text{(since } \eta(\tilde{x}(t,\epsilon)) \in \text{dom } H_t) \\
&= <\check{H} \tilde{\eta}(\tilde{x}), \tilde{\eta}(\tilde{y})>,
\end{aligned}$$

and the proof is complete. □

We shall now head towards a "usable" description of $S(\widetilde{M})$. The setting -- for the rest of this section -- is as above: $(M,G,\alpha)$ is a discrete dynamical system; $M$ is semifinite; $\phi$ is a fns trace on $M$ and $\widetilde{\phi}$ the induced fns weight on $\widetilde{M}$, given by $\widetilde{\phi}(\widetilde{x}) = \phi(\widetilde{x}(\epsilon,\epsilon))$; $M$ is "standard relative to $\phi$" (i.e., $M \subseteq \mathfrak{L}(\mathcal{H})$, $\mathcal{H} = \mathcal{H}_\phi$, $\pi_\phi = id_M$ and $\eta$ (= $\eta_\phi$): $N$ (= $N_\phi$) $\to \mathcal{H}$) so that $\widetilde{M}$ is standard relative to $\widetilde{\phi}$ (by Prop. 4.2.2) with $\widetilde{\eta}$ (= $\eta_{\widetilde{\phi}}$): $\widetilde{N}$ (= $N_{\widetilde{\phi}}$) $\to \widetilde{\mathcal{H}}$ given by $(\widetilde{\eta}(\widetilde{x}))(s) = \eta(\widetilde{x}(s,\epsilon))$; for $t$ in $G$, $\phi \circ \alpha_t = \phi(H_t \cdot)$, where $H_t$ is a (uniquely determined) positive invertible self-adjoint operator affiliated to $Z(M)$; finally

$$\widetilde{\Delta} = \Delta_{\widetilde{\phi}} = \bigoplus_{t \in G} H_t.$$

We shall also assume henceforth that the action $\alpha$ is free and ergodic, so that -- by Prop. 4.1.15 -- $\widetilde{M}$ is a factor.

**Lemma 4.2.6.**

(a) $\pi_\alpha(M) \subseteq \widetilde{M}^{\widetilde{\phi}}$;

(b) $Z(\widetilde{M}^{\widetilde{\phi}}) \subseteq \pi_\alpha(Z(M))$.

**Proof.** If $\widetilde{x} = \pi(x) \in \pi(M)$, then by Prop. 4.2.5,

$$\sigma_t^{\widetilde{\phi}}(\widetilde{x}) = \bigoplus_s H_s^{it}\, \alpha_{s^{-1}}(x) H_s^{-it} = \pi(x),$$

since $H_s\ \eta\ Z(M)$ for all $s$, and (a) is proved.

If $\widetilde{x} \in Z(\widetilde{M}^{\widetilde{\phi}})$, then by (a) above, $\widetilde{x} \in \widetilde{M} \cap \pi_\alpha(M)'$; the assumption that $\alpha$ is free, together with Corollary 4.1.9, completes the proof of (b). □

**Lemma 4.2.7.** *Let* $0 \neq e \in P(Z(M))$ *and* $\widetilde{e} = \pi(e)$ *(so that* $0 \neq \widetilde{e} \in P(\widetilde{M}^{\widetilde{\phi}})$ *by Lemma 4.2.6 (a));*

(a) *let* $\widetilde{x} \in \widetilde{M}$*; then*

$$\widetilde{x} \in \widetilde{M}_{\widetilde{e}} \Leftrightarrow \widetilde{x}(s, \epsilon) = \alpha_{s^{-1}}(e)e\, \widetilde{x}(s,\epsilon), \quad \textit{for all } s \textit{ in } G\ ;$$

(b) $\sigma_t^{\widetilde{\phi}}(\widetilde{M}_{\widetilde{e}}) = \widetilde{M}_{\widetilde{e}}$ *for all* $t$ *in* $\mathbb{R}$ *;*

(c) *if* $K_e = [\widetilde{\eta}(\widetilde{N} \cap \widetilde{M}_{\widetilde{e}})]$ *and* $\widetilde{p}_e = p_{K_e}$*, then*

$$\widetilde{p}_e = \bigoplus_{s \in G} (\alpha_{s^{-1}}(e)e).$$

**Proof.**

(a) $\tilde{x} = \tilde{e}\,\tilde{x}\,\tilde{e} \Leftrightarrow \tilde{x}(s, \epsilon) = (\tilde{e}\,\tilde{x}\,\tilde{e})(s, \epsilon) \quad \forall s \text{ in } G$

$$\Leftrightarrow \tilde{x}(s, \epsilon) = \alpha_{s^{-1}}(e)\tilde{x}(s, \epsilon)e \quad \forall s \text{ in } G;$$

notice that $\check{e} \in Z(M)$.

(b) If $\tilde{x} \in \check{M}_{\tilde{e}}$ and $t \in \mathbb{R}$,

$$(\sigma_t^{\check{\phi}}(\tilde{x}))(s, \epsilon) = (\check{\Delta}^{it}\,\tilde{x}\,\check{\Delta}^{-it})(s, \epsilon)$$

$$= H_s^{it}\,\tilde{x}(s, \epsilon)H_\epsilon^{-it}$$

$$= H_s^{it}\,\tilde{x}(s, \epsilon) = \alpha_{s^{-1}}(e)e\,H_s^{it}\,\tilde{x}(s, \epsilon)$$

since $x(s, \epsilon) \in M_{\alpha_{s^{-1}}(e)e}$, by (a), and $H_s^{it} \in Z(M)$; so

$$(\sigma_t^{\check{\phi}}(\tilde{x}))(s,\epsilon) = \alpha_{s^{-1}}(e)e(\sigma_t^{\check{\phi}}(\tilde{x}))(s,\epsilon)$$

and so, by (a), $\sigma_t^{\check{\phi}}(\check{M}_{\tilde{e}}) \subseteq \check{M}_{\tilde{e}}$; consider $(-t)$ rather than $t$, and conclude that the above inclusion may be replaced by equality.

(c) Define

$$\tilde{p} = \underset{s \in G}{\oplus}(\alpha_{s^{-1}}(e)e),$$

and observe that $\tilde{p}$ is a projection (since $e \in Z(M)$); the definition of $\check{\eta}$, together with (a) above, shows that $\tilde{p}$ fixes every vector in a dense subset of $\mathcal{K}_e$, and hence $\mathcal{K}_e \subseteq \operatorname{ran} \tilde{p}$. On the other hand, it is easy to see (again using (a)) that

$$\tilde{p}(\check{\eta}(\check{N} \cap \check{M}_0)) \subseteq \check{\eta}(\check{N} \cap \check{M}_e \cap \check{M}_0) \subseteq \mathcal{K}_e,$$

where, as usual, $\check{M}_0$ is as in Ex. (4.1.3) (c); since $\check{\eta}(\check{N} \cap \check{M}_0)$ is dense in $\check{\mathcal{H}}$, conclude that ran $\tilde{p} \subseteq \mathcal{K}_e$; hence $\tilde{p} = \tilde{p}_e$, as desired. □

**Lemma 4.2.8.** *Let* $e$, $\tilde{e}$, $\mathcal{K}_e$, $\tilde{p}_e$ *be as in Lemma* 4.2.7. *Then* $\tilde{p}_e\check{\Delta} \subseteq \check{\Delta}\,\tilde{p}_{\tilde{e}}$, *and* $\Delta_{\check{\Phi}_{\tilde{e}}}$ *may be identified with the restriction of* $\oplus_{t \in G} H_t$ *to* $\mathcal{K}_e$.

**Proof.** Since

$$\tilde{p}_e = \underset{t \in G}{\oplus}(\alpha_{t^{-1}}(e)e) \quad \text{and} \quad \check{\Delta} = \underset{t \in G}{\oplus} H_t\,,$$

and since $H_t\ \eta\ Z(M)$, $e$, $\alpha_t\text{-}1(e) \in Z(M)$, it is clear that $\tilde{p}_e\,\check{\Delta} \subseteq \check{\Delta}\,\tilde{p}_e$. It is easy to verify now that

$$(\ \mathcal{K}_e, id_{\check{M}_{\tilde{e}}}, \check{\eta}|N_{\check{\Phi}_{\tilde{e}}}\ )$$

is a GNS triple for $(\check{M}_{\tilde{e}},\check{\Phi}_{\tilde{e}})$; an argument similar to the one used in proving Proposition 2.6.6 proves that

$$\Delta_{\tilde{\phi}_{\tilde{e}}} = \hat{\Delta}| K_e. \qquad \square$$

**Proposition 4.2.9.** *Let $\alpha$ be a free, ergodic action of a discrete group $G$ on a semifinite von Neumann algebra. The following conditions on a non-negative real number $\lambda$ are equivalent*:

(i) $\lambda \in S(M \otimes_\alpha G)$;
(ii) *for every $\epsilon > 0$ and non-zero $e$ in $P(Z(M))$, there exist $t$ in $G$ and non-zero $f$ in $P(Z(M))$ such that $f \vee \alpha_t(f) \leq e$, and* $\text{sp}(H_t f | \text{ran } f) \subseteq (\lambda - \epsilon, \lambda + \epsilon)$.

**Proof.** Let us retain the notation established so far in this section.

Since, by Lemma 4.2.6, $Z(\tilde{M}^{\tilde{\phi}}) \subseteq \pi(Z(M)) \subseteq \tilde{M}^{\tilde{\phi}}$, it follows from Corollary 3.4.8 and Ex. (3.4.9) (b) that

$$S(\tilde{M}) = \cap \{\text{sp } \Delta_{\tilde{\phi}_{\tilde{e}}} : 0 \neq \tilde{e} \in \pi(P(Z(M))\}.$$

For notational convenience, let us write $K(t,e) = \text{ran}(\alpha_{t^{-1}}(e)e)$ and $H(t,e) = H_t|(K(t,e) \cap \text{dom } H_t)$, whenever $t \in G$ and $0 \neq e \in P(Z(M))$. Notice that all the operators in sight commute and hence $H(t,e)$ is a self-adjoint operator in $K(t,e)$. It follows from Lemma 4.2.8 that $\Delta_{\tilde{\phi}_{\tilde{e}}}$ (where $\tilde{e} = \pi(e)$, $e \in P(Z(M))$ ) may be identified with $\oplus_{t \in G} H(t,e)$, so that

$$\text{sp } \Delta_{\tilde{\phi}_{\tilde{e}}} = \left[ \bigcup_{t \in G} \text{sp } H(t,e) \right]^- .$$

So, if $\lambda \in [0,\infty)$, conclude that

$$\lambda \in S(\tilde{M}) \Leftrightarrow \lambda \in \text{sp } \Delta_{\tilde{\phi}_{\tilde{e}}} \text{ whenever } 0 \neq \tilde{e} = \pi(e), e \in P(Z(M))$$

$$\Leftrightarrow \forall\, 0 \neq e \in P(Z(M)) \text{ and } \forall \epsilon > 0, \exists t \text{ in } G$$

$$\text{such that } f = 1_{(\lambda-\epsilon, \lambda+\epsilon)}(H(t,e)) \neq 0. \qquad \square$$

Before concluding this section, let us consider a special case, and for that case, reformulate Proposition 4.2.9 in a form that will be tailor-made for use in the next section.

Let $M = L^\infty(X,\mathcal{F},\mu)$ with $(X,\mathcal{F},\mu)$ a separable and $\sigma$-finite measure space; let $t \to T_t$ be a homomorphism from $G$ into the group of automorphisms of $(X,\mathcal{F},\mu)$ (cf. Definition 4.1.10), and let $t \to \alpha_t$ be the induced action of $G$ on $M$: $\alpha_t(f) = f \circ T_t^{-1}$; let $\phi$ be the fns trace on $M$ given by $\phi(f) = \int f \, d\mu$; if $E \in \mathcal{F}$,

$$\phi \circ \alpha_t(1_E) = \phi(1_{T_t(E)}) = \mu(T_t(E)),$$

and so $\phi \circ \alpha_t = \phi(H_t \cdot)$, where $H_t$ is the operator in $L^2(X,\mathcal{F},\mu)$ of multiplication by $d\mu o T_t/d\mu$.

It is easy, now, to deduce from Proposition 4.2.9, that if the action $\{T_t\}$ of $G$ on $(X,\mathcal{F},\mu)$ is free and ergodic (in the sense of Examples 4.1.11 and 4.1.14), and if $\mathcal{G} = \{T_t: t \in G\}$, then $S(L^\infty(X,\mathcal{F},\mu) \otimes_\alpha G) = r(\mathcal{G})$, where $r(\mathcal{G})$ is Krieger's notion of the ratio set of a group $\mathcal{G}$ of automorphisms of $(X,\mathcal{F},\mu)$ (cf. [Kri]) defined below.

**Definition 4.2.10.** Let $\mathcal{G}$ be a countable group of automorphisms of $(X,\mathcal{F},\mu)$. Define the ratio set $r(\mathcal{G})$ thus: $r(\mathcal{G}) \subseteq [0,\infty)$; if $\lambda \geq 0$, then $\lambda \in r(\mathcal{G})$ if and only if, for every $\epsilon > 0$ and $E \in \mathcal{F}$ such that $\mu(E) > 0$, there exist $T \in \mathcal{G}$ and $F \in \mathcal{F}$ such that $\mu(F) > 0$, $F \cup T(F) \subseteq E$ and $|(d\mu o T/d\mu)(\omega) - \lambda| < \epsilon$ for all $\omega$ in $F$. □

## Exercises

**(4.2.11)** Let $(M,G,\alpha)$, $\phi$, $H_t$, $\breve{\phi}$ be as in the second half of this section.

(a) If $\tilde{x} \in \breve{M}$, show that

$$\breve{\phi}(\tilde{x}^*\tilde{x}) = \sum_{s\in G} \phi(\tilde{x}(s, \epsilon)^* \tilde{x}(s, \epsilon))$$

and

$$\breve{\phi}(\tilde{x}\tilde{x}^*) = \sum_{s\in G} \phi \circ \alpha_s(\tilde{x}(s, \epsilon)\tilde{x}(s, \epsilon)^*);$$

deduce that $\breve{\phi}$ is a trace if and only if $\phi$ is $G$-invariant: i.e., $\phi \circ \alpha_t = \phi \ \forall t$ in $G$.

(b) If $\phi$ is $G$-invariant, deduce the semifiniteness of $\breve{M}$ by showing directly that $S(\breve{M}) = \{1\}$. (Hint: use Prop. 4.2.9.)

**(4.2.12)** Let the notation and hypothesis be as above. For each $t$ in $G$, let $e_t = 1_{\{1\}}(H_t)$.

(a) Show that $\breve{M}^{\breve{\phi}} = \{\tilde{x} \in \breve{M}: \tilde{x}(s, \epsilon) \in M_{e_s} \ \forall s \text{ in } G\}$.

(b) Show that $\breve{M}^{\breve{\phi}} = \pi_\alpha(M)$ if and only if $e_s = 0$ for $s \neq \epsilon$.

**(4.2.13)** Let $M$ be a semifinite factor with (essentially uniquely determined) fns trace $\tau$. Let $\theta$ be an automorphism of $M$ such that $\tau \circ \theta = \lambda\tau$ for some $\lambda$ in $(0,1)$.

(a) Show that $\theta$ is free in the sense of Definition 4.1.8 (a). (Hint: if $\theta$ is inner, the traciality of $\tau$ would force $\tau \circ \theta = \tau$.)

(b) Define an action $\alpha$ of $\mathbb{Z}$ on $M$ by $\alpha_n = \theta^n$. Show that $\breve{M}$ is a factor. (Hint: in fact, from (a), deduce that $M \cap \pi(M)' = \{\lambda.1: \lambda \in \mathbb{C}\}$.)

(c) $M$ is of type $\mathrm{III}_\lambda$. (Hint: use Prop. 4.2.9.)

(d) Assume the fact that every automorphism of a factor of type I is inner, and show that a factor $M$ as above, is necessarily of type $\mathrm{II}_\infty$.

**(4.2.14)** Let $M$ be a semifinite factor with fns trace $\tau$. Suppose $\alpha$ is an action of $Q$ -- the additive group of rational numbers -- on $M$ such that $\tau \circ \alpha_t = e^t\tau$ for all $t$ in $Q$. Show that $M \otimes_\alpha Q$ is a factor of type $\mathrm{III}_1$. (Hint: as in Ex. (4.2.13), show that the action $\alpha$ is free, deduce that $\tilde{M}$ is a factor, notice that $H_t = e^t 1$, and use Prop. 4.2.9.)

## 4.3. Examples of Factors

If $\mathcal{H}$ is a separable Hilbert space of dimension $n$ ($1 \leqslant n \leqslant \infty$) and $M = \mathcal{L}(\mathcal{H})$, then $M$ is clearly a factor. The function $D$, defined by $D(\mathcal{M}) = \dim \mathcal{M}$, is a dimension function for $M$ and consequently $M$ is of type $\mathrm{I}_n$. The proof of the converse assertion -- that any factor of type $\mathrm{I}_n$ is isomorphic to $\mathcal{L}(\mathcal{H})$ for an $n$-dimensional Hilbert space $\mathcal{H}$ -- is outlined in the following exercises.

### Exercises

**(4.3.1)** Let $M$ be a factor of type $\mathrm{I}_n$, $1 \leqslant n \leqslant \infty$.

(a) If $e$ is a minimal projection in $M$, then $eMe = \{\lambda e: \lambda \in \mathbb{C}\}$. (Hint: if $x = x^* \in M$, consider the spectral projections of $exe$.)

(b) Any two minimal projections are equivalent; further, if $e$ and $f$ are minimal projections and if $u$ and $v$ are partial isometries in $M$ such that $u^*u = v^*v = e$ and $uu^* = vv^* = f$, then there exists a complex scalar $\lambda$ of unit modulus such that $u = \lambda v$. (Hint: consider $u^*v$ and use (a).)

(c) There exists a family $\{e_i: i \in I\}$ of pairwise orthogonal minimal projections in $M$ such that $1 = \bigvee_{i\in I} e_i$, where $I = \{1,2, ..., n\}$ or $\{1,2, ...\}$ according as $n$ is finite or infinite.

(d) With $\{e_i\}$ as in (c), pick partial isometries $u_i$ in $M$ such that $u_i^*u_i = e_1$, $u_iu_i^* = e_i$; for any $x$ in $M$, $i,j$ in $I$, show that $e_ixe_j = \lambda_{ij}u_iu_j^*$ for some $\lambda_{ij}$ in $\mathbb{C}$. (Hint: let $e_ixe_j = uh$ be the polar decomposition of $e_ixe_j$; apply (b) and (a) to $u$ and $h$ respectively.)

(e) Suppose $M \subseteq \mathcal{L}(\mathcal{H})$. Let $\mathcal{M} = \operatorname{ran} e_1$ and let $\mathcal{H}_n$ be an $n$-dimensional Hilbert space with a fixed orthonormal basis $\{\xi_i: i \in I\}$. Show that there exists a unique unitary operator $u: \mathcal{H}_n \otimes \mathcal{M} \to \mathcal{H}$ such that $u(\xi_i \otimes \eta) = u_i\eta$, for $i \in I$, $\eta \in \mathcal{M}$.

(f) With $u$ as in (e), show that

$$u^*Mu = \{x \otimes 1 \in \mathcal{L}(\mathcal{H}_n \otimes \mathcal{M}): x \in \mathcal{L}(H_n)\}.$$

(Hint: use (d).) □

We turn now to factors of type II.

**Proposition 4.3.2.** *Let $G$ be a countable discrete group, with group von Neumann algebra $W^*(G) = M$ (cf. Example 2.3.7 (c)).*

(a) *A necessary and sufficient condition for $W^*(G)$ to be a factor is that for every $t \neq \epsilon$ in $G$, the conjugacy class $\{sts^{-1}: s \in G\}$ is infinite.*
(b) *If the "infinite conjugacy class condition" is satisfied, then $W^*(G)$ is a factor of type $II_1$, unless $G = \{\epsilon\}$.*

**Proof.** If $\tilde{x} \in \breve{M} = W^*(G)$ -- recall from Example 4.1.4 (a) that $W^*(G) = \mathbb{C} \otimes_\alpha G$, where $\alpha$ is the trivial action of $G$ on $\mathbb{C}$ -- let $((\tilde{x}(s,t)))$ denote the matrix of $\tilde{x}$ with respect to canonical orthonormal basis $\{\xi_t: t \in G\}$ of $\ell^2(G)$. Then $\tilde{x}(s,t) = \tilde{x}(st^{-1},\epsilon)$ for all $s,t$ in $G$; further, it follows from Lemma 4.1.7 (b) that $\tilde{x} \in Z(M) \Leftrightarrow \tilde{x}(sts^{-1},\epsilon) = \tilde{x}(t,\epsilon)$ for all $s,t$ in $G$.

Also, $\Sigma_s|\tilde{x}(s,\epsilon)|^2 = \|\tilde{x}\ \xi_\epsilon\|^2 < \infty$; hence if every conjugacy class other than $\{\epsilon\}$ is infinite, it follows that $\tilde{x} \in Z(M) \Leftrightarrow \tilde{x}(s,\epsilon) = 0$ for $s \neq \epsilon$ $\Leftrightarrow \tilde{x} = \tilde{x}(\epsilon,\epsilon)1$. On the other hand, if $\{sts^{-1}: s \in G\}$ is finite for some $t \neq \epsilon$, define $\tilde{x} \in M_0$ by requiring that

$$\tilde{x}(s,\ \epsilon) = \begin{cases} 1, & \text{if } s = utu^{-1} \text{ for some } u \text{ in } G \\ 0, & \text{otherwise,} \end{cases}$$

and note that $\tilde{x}$ is a non-scalar central element of $\breve{M}$.

Suppose now that $M$ is a factor. Since the equation $\phi(\lambda) = \lambda$ defines a faithful normal state $\phi$ on $\mathbb{C}$, it follows that the equation $\breve{\phi}(\tilde{x}) = \tilde{x}(\epsilon,\epsilon)$ defines a faithful normal state on $\breve{M}$. If $\tilde{x} \in \breve{M}$,

$$\breve{\phi}(\tilde{x}^*\tilde{x}) = \sum_s |\tilde{x}(s,\ \epsilon)|^2,$$

while

$$\begin{aligned} \breve{\phi}(\tilde{x}\tilde{x}^*) &= \Sigma\ |\breve{x}^*(s, \epsilon)|^2 \\ &= \Sigma\ |\tilde{x}(\epsilon, s)|^2 \\ &= \Sigma\ |\tilde{x}(s^{-1}, \epsilon)|^2, \end{aligned}$$

and consequently $\breve{\phi}$ is a trace on $\breve{M}$; it follows that $\breve{M}$ is a factor of finite type. If $G \neq \{\epsilon\}$, then $G$ is infinite (by (a)); the set $\{\lambda_t: t \in G\}$ is clearly linearly independent (look at their matrices!) and so $W^*(G)$ is not finite-dimensional, and consequently $W^*(G)$ cannot be of type $I_n$ for a finite $n$. The only remaining possibility is that $W^*(G)$ is of type $II_1$. □

## Exercises

**(4.3.3)** Verify that the following countable groups satisfy the "infinite conjugacy class condition" of Prop. 4.3.2 (a):

(a) the group of permutations of $\mathbb{N} = \{1,2,...,\}$ which move only finitely many integers;
(b) a finitely generated free group on two or more generators. □

Next in line are factors of type $\mathrm{II}_\infty$. We shall show that if $M$ is a factor of type $\mathrm{II}_1$ and if $\mathcal{H}$ is a separable infinite-dimensional Hilbert space, then $M \otimes \mathcal{L}(\mathcal{H})$ is a factor of type $\mathrm{II}_\infty$, and that conversely every factor of type $\mathrm{II}_\infty$ arises in this fashion.

If $\mathcal{H}$ is a separable Hilbert space, let

$$\overline{\mathcal{H}} = \bigoplus_{n=1}^{\infty} \mathcal{H}_n ,$$

where $\mathcal{H}_n = \mathcal{H}$ for all $n$. Then, as in our discussion of discrete crossed-products, we shall identify an operator $\overline{x}$ on $\overline{\mathcal{H}}$ with a matrix $((\overline{x}(m,n)))_{m,n=1}^{\infty}$ where $\overline{x}(m,n) \in \mathcal{L}(\mathcal{H})$ for all $m$ and $n$. If $M$ is a von Neumann algebra of operators on $\mathcal{H}$, let $\overline{M} = \{\overline{x} \in \mathcal{L}(\overline{\mathcal{H}}): \overline{x}(m,n) \in M\ \forall m,n\}$; it is clear that $\overline{M}$ is a von Neumann algebra of operators on $\overline{\mathcal{H}}$.

**Proposition 4.3.4.** *Let $M \subseteq \mathcal{L}(\mathcal{H})$, $\overline{M} \subseteq \mathcal{L}(\overline{\mathcal{H}})$ be as above.*

(a) *If $M$ is a factor of type $\mathrm{II}_1$, then $\overline{M}$ is a factor of type $\mathrm{II}_\infty$;*
(b) *If $\hat{M}$ is a factor of type $\mathrm{II}_\infty$ operating on a Hilbert space $\hat{\mathcal{H}}$, there exists a factor $M$ of type $\mathrm{II}_1$ acting on a Hilbert space $\mathcal{H}$ and a unitary operator $u$: $\overline{\mathcal{H}} \to \hat{\mathcal{H}}$ such that $u\overline{M}u^* = \hat{M}$.*

**Proof.** (a) It is not hard to show that if $\pi$: $M \to N$ is an isomorphism of von Neumann algebras $M \subseteq \mathcal{L}(\mathcal{H})$ and $N \subseteq \mathcal{L}(\mathcal{K})$, then the von Neumann algebras $\overline{M}$ and $\overline{N}$ are isomorphic; the details are outlined in Exercises (4.3.5) and (4.3.6). Hence, we may assume that $M$ is standard relative to a faithful normal tracial state $\tau$ on $M$; thus, assume that $\Omega$ is a unit vector in $\mathcal{H}$ which is cyclic and separating for $M$ and that the equation $\tau(x) = \langle x\Omega,\Omega\rangle$ defines a faithful normal tracial state on $M$; thus $\|x\Omega\| = \|x^*\Omega\|$ for all $x$ in $M$.

Define $\overline{\tau}$: $\overline{M}_+ \to [0,\infty]$ by

$$\overline{\tau}(\overline{x}) = \sum_{n=1}^{\infty} \langle \overline{x}(n,n)\Omega,\Omega\rangle, \quad \overline{x} \in \overline{M}_+ .$$

It is clear that $\overline{\tau}$ is a weight on $\overline{M}$, which is faithful and normal. (Reason: if $\overline{x} \geq 0$ and $\overline{x}(n,n) = 0$ for all $n$, then $\overline{x} = 0$, so $\overline{\tau}$ is faithful; if

$$\psi_n(\overline{x}) = \sum_{k=1}^{n} \langle \overline{x}(k,k)\Omega,\Omega\rangle ,$$

then each $\psi_n$ is a normal positive linear functional on $\overline{M}$ and $\psi_k \nearrow \overline{\tau}$, so that, by Prop. 2.4.9, $\bar{\tau}$ is normal.) If $\bar{M}_0 = \{\bar{x} \in \bar{M}: \bar{x}(m,n) = 0$ for all but finitely many pairs $(m,n)\}$, it is easy to see that $\bar{M}_0$ is a self-adjoint subalgebra which is $\sigma$-strongly* dense in $\bar{M}$; since $\bar{M}_0$ is readily verified to be a subset of $\mathcal{M}_{\bar{\tau}}$ it is seen that $\bar{\tau}$ is semifinite.

Finally, if $\bar{x} \in \bar{M}$, notice that

$$\begin{aligned}\bar{\tau}(\bar{x}\,\bar{x}^*) &= \sum_{m,n} \|\bar{x}^*(m,n)\Omega\|^2 \\ &= \sum_{m,n} \|\bar{x}(n,m)^*\Omega\|^2 \\ &= \sum_{m,n} \|\bar{x}(n,m)\Omega\|^2 \quad \text{(since } \tau \text{ is a trace)} \\ &= \bar{\tau}(\bar{x}^*\,\bar{x})\end{aligned}$$

and consequently $\bar{\tau}$ is a trace; hence $\bar{M}$ is semifinite.

It is easily checked that (even if $M$ is not a factor) $\bar{M}' = \{\bar{x}' \in \mathcal{L}(\bar{\mathcal{H}}): \bar{x}'(m,n) = \delta_{mn}x'$ for all $m$ and $n$, for some $x' \in M'\}$ (as in the proof of the double commutant theorem); hence $\bar{M}$ inherits factoriality from $M$.

From the foregoing discussion $\bar{M}$ is seen to be a semifinite factor which is not finite, since

$$\bar{\tau}(1) = \sum_{n=1}^{\infty} \|\Omega\|^2 = \infty ;$$

hence $\bar{M}$ is of type $I_\infty$ or $II_\infty$. However, since $M$ is of type $II_1$, there exist non-zero projections in $M$ with arbitrarily small "dimension"; it follows that $\bar{M}$ also has such projections. (Put $\bar{e}(m,n) = \delta_{m1}\delta_{n1}e$, where $0 < \tau(e) < \epsilon$). Thus $\bar{M}$ cannot be of type I.

(b) Let $\hat{e}$ be a non-zero finite projection in $\hat{M}$. It is quite easy to see that there exists a sequence $\{e_n\}_{n=1}^{\infty}$ of pairwise orthogonal projections in $\hat{M}$ such that $1 = \Sigma e_n$ and $e_n \sim \hat{e}$ for all $n$. (Apply Proposition 1.2.3 with $\mathcal{M} = \hat{\mathcal{H}}$, $\mathcal{N} = \operatorname{ran} \hat{e}$, note that $\hat{\mathcal{H}}$ is infinite since $\hat{M}$ is of type $II_\infty$, and appeal to Theorem 1.2.10). Fix $u_n$ in $\hat{M}$ such that $u_n^*u_n = \hat{e}$ and $u_nu_n^* = e_n$ for each $n$.

Let $\mathcal{H} = \operatorname{ran} \hat{e}$ and $M = \hat{e}\hat{M}\hat{e}$, viewed as a von Neumann algebra acting on $\mathcal{H}$. It is a fact -- which we will not go into, here, and which the reader may find in [Dix] -- that $M$ inherits the property of being a factor from $\hat{M}$. If $\hat{\tau}$ is a fns trace on $\hat{M}$, observe that $\hat{\tau}|\, P(M)$ is a dimension function for $M$; since $\hat{e}$ is the identity element of $M$, and since $\hat{e}$ is finite, it follows that $M$ is a finite factor. Since $\hat{M}$ does not contain minimal projections, $M$ cannot contain minimal projections and so $M$ is not of type I. The only remaining possibility is that $M$ is a factor of type $II_1$.

With $\mathcal{H}$ and $M$ as in (a), it is easy to see that the equation

$$u\bar{\xi} = \sum_{n=1}^{\infty} u_n\bar{\xi}(n)$$

defines a unitary operator $u: \bar{\mathcal{H}} \to \hat{\mathcal{H}}$. With $\bar{M}_0$ as in the proof of (a), an easy computation reveals that if $\bar{x} \in \bar{M}_0$, then

$$u\, \bar{x}\, u^* = \sum_{m,n=1}^{\infty} u_m\, \bar{x}(m,n) u_n^* ;$$

since sums of this sort are easily verified to constitute a $\sigma$-weakly dense self-adjoint subalgebra of $\hat{M}$, conclude that $u\, \bar{M}\, u^* = \hat{M}$. □

### Exercises

**(4.3.5)**

(a) If $\pi: M \to N$ is a *-algebra isomorphism of a von Neumann algebra $M$ onto a von Neumann algebra $N$, then $\pi$ is a $\sigma$-weak homeomorphism. (Hint: $\pi(M_+) \subseteq \pi(N_+)$ and so $\pi$ preserves order; since the same holds for $\pi^{-1}$, if $x,y \in M_+$, then $x \leqslant y \Leftrightarrow \pi(x) \leqslant \pi(y)$; conclude that $\pi$ is normal and hence $\sigma$-weakly continuous -- cf. the proof of Theorem 2.2.1; the same reasoning holds for $\pi^{-1}$.).

Let $M$ ($\subseteq \mathfrak{L}(\mathcal{H})$) and $N$ ($\subseteq \mathfrak{L}(K)$) be von Neumann algebras, and suppose $\pi: M \to N$ is a (necessarily normal) *-isomorphism of $M$ onto $N$. Let $\bar{\mathcal{H}}$, $\bar{M}$, $\bar{K}$, $\bar{N}$ be constructed as in Prop. 4.3.4, and let $\bar{M}_0$ and $\bar{N}_0$ be as in the proof of Prop. 4.3.4.

(b) If $\bar{x} = ((\bar{x}(m,n))) \in \mathfrak{L}(\bar{\mathcal{H}})$ and $k = 1,2, \ldots,$ define $\bar{x}_k \in \mathfrak{L}(\bar{\mathcal{H}})$ by

$$\bar{x}_k(m,n) = \begin{cases} \bar{x}(m,n), & \text{if } m,n \leqslant k \\ 0, & \text{otherwise.} \end{cases}$$

Show that $\|\bar{x}_k\| \leqslant \|\bar{x}\|$ for all $k$ and $\bar{x}_k \to \bar{x}$ $\sigma$-strongly*.

(c) If $\bar{x}(m,n) \in \mathfrak{L}(\mathcal{H})$ for $m,n = 1,2, \ldots,$ define $\bar{x}_k \in \mathfrak{L}(\bar{\mathcal{H}})$ as in (a) above. Show that the matrix $((\bar{x}(m,n)))$ represents a bounded operator on $\bar{\mathcal{H}}$ iff $\sup_k \|\bar{x}_k\| < \infty$.

(d) For $k = 1,2, \ldots,$ let $\bar{M}_k = \{\bar{x} \in \bar{M} : \bar{x}(m,n) \neq 0 \Rightarrow m,n \leqslant k\}$, and let $\bar{N}_k$ be similarly defined. Then $\bar{M}_k$ (resp. $\bar{N}_k$) may be regarded as a von Neumann algebra of operators on $\mathcal{H}^{(k)}$ (resp., $K^{(k)}$), where $\mathcal{H}^{(k)}$ is the direct sum of $k$ copies of $\mathcal{H}$. Show that there is a unique *-isomorphism $\bar{\pi}: \bar{M}_k \to \bar{N}_k$ such that $\bar{\pi}(\bar{x}) = ((\pi(\bar{x}(m,n))))$; further $\bar{\pi}$ is isometric and a $\sigma$-weak homeomorphism. (Hint: use (a) and the fact that an injective *-homomorphism is isometric -- cf. second half of proof of Theorem 2.2.1.)

(e) Show that there exists a unique (normal and isometric) *-isomorphism $\bar{\pi}$ of $\bar{M}$ onto $\bar{N}$ such that $\bar{\pi}(\bar{x}) = ((\pi(\bar{x}(m,n)))$. (Hint: use (d), (c) and (b) above.) □

Recall that a factor $\tilde{M}$ is of type $\mathrm{III}_0$, $\mathrm{III}_\lambda$ $(0 < \lambda < 1)$ or $\mathrm{III}_1$ according as $S(M)$ is $\{0,1\}$, $\{0\} \cup \{\lambda^n : n \in \mathbb{Z}\}$ or $[0,\infty)$ (cf. the discussion following Prop. 3.4.6). So, in view of Proposition 4.2.9, in order to construct an example of a factor of type $III_0$, $III_\lambda$ $(0 < \lambda < 1)$ or $III_1$, it is sufficient to construct an example of a countable group $G$ of automorphisms of a separable and $\sigma$-finite measure space $(X,\mathcal{F},\mu)$, which acts freely and ergodically, and has ratio set given by $r(G) = \{0,1\}$, $\{0\} \cup \{\lambda^n: n \in \mathbb{Z}\}$ or $[0,\infty)$ (cf. Definition 4.2.10 and the preceding discussion).

Let $X_0 = \{1,2, ..., N\}$ be a finite set and let $\mu_0$ be a probability measure defined on the subsets of $X_0$ such that $\mu_0(\{j\}) = p_j > 0$ for $1 \leq j \leq N$ (and $\Sigma p_j = 1$). Let $X = X_0^{\mathbb{N}} = \{\omega: \mathbb{N} \to X_0\}$, where $\mathbb{N} = \{1,2, ...\}$. Equip $X$ with the product $\sigma$-algebra $\mathcal{F}$ and the product measure

$$\mu = \bigotimes_{n=1}^{\infty} \mu_n ,$$

with $\mu_n = \mu_0$ for all $n$. By a cylinder set in $X$, we shall mean a set of the form $\{\omega \in X: (\omega(1), ..., \omega(n)) \in E_n\}$ where $E_n$ is any subset of

$$X_0^n = \underbrace{X_0 \times \cdots \times X_0}_{n} ,$$

and $n = 1,2, ...$; thus $\mathcal{F}$ is the $\sigma$-algebra generated by cylinder sets. By an elementary cylinder set in $X$, we shall mean a set of the form $\{\omega \in X: \omega(n) = j_0\}$ for some $n$ in $\mathbb{N}$ and $j_0$ in $X_0$.

For each permutation $\sigma$ of $\{1, ..., N\}$ and $k$ in $\mathbb{N}$, let $T_{\sigma,k}: X \to X$ be defined by

$$(T_{\sigma,k}\omega)(m) = \begin{cases} \omega(m), & \text{if } m \neq k \\ \sigma(\omega(k)), & \text{if } m = k \end{cases} .$$

Since $p_j > 0$ for each $j$, it is clear that each $T_{\sigma,k}$ is an automorphism of $(X,\mathcal{F},\mu)$; let $G$ be the group generated by $\{T_{\sigma,k}: \sigma \in C_n, k \in \mathbb{N}\}$, where $C_N$ is the cyclic subgroup of $S_N$ generated by a full cycle, say $(1\ 2\ ...\ N)$.

There are several ways to see that $G$ acts freely and ergodically on $(X,\mathcal{F},\mu)$, one of which is outlined in the following exercises.

## Exercises

**(4.3.6)** For $j = 1, ..., N$ and $k = 1,2, ...$, let $C_{j,k}$ be the elementary cylinder set $\{\omega \in X: \omega(k) = j\}$.

(a) If $\omega_1,\omega_2 \in X$, show that $\omega_1 = \omega_2$ if and only if

$$1_{C_{j,k}}(\omega_1) = 1_{C_{j,k}}(\omega_2),$$

for $j$ in $X_0$, $k$ in $\mathbb{N}$ .

(b) For each $\sigma \in C_n$, $\sigma \neq \epsilon$ and $k \in \mathbb{N}$, show that $T_{\sigma,k}\omega \neq \omega$ for all $\omega$ in $X$.

(c) If $T_{\sigma,k}$ is as in (b), show that $T_{\sigma,k}$ acts freely. (Hint: use (a) above and Ex. (4.1.12).)

**(4.3.7)** (The aim of this exercise is to establish that $G$ acts ergodically on $X$; the reader who knows some probability theory will realize that this is a special case of Kolmogorov's zero-one law.) Suppose $E \in \mathcal{F}$ and $\mu(TE \Delta E) = 0$ for all $T$ in $G$.

(a) Let $F = \cup_{T\in G} TE$; then $F \in \mathcal{F}$, $\mu(F) = \mu(E)$ and $TF = F$ for all $T$ in $G$. (Hint: $G$ is countable.)

(b) If, for $k = 1,2, \ldots$ $\pi_{k]} : X \to X_0^{\{1,2,\ldots,k\}}$ and $\pi_{(k} : X \to X_0^{\{k+1,k+2,\ldots\}}$ denote the natural projections (so that $\omega = (\pi_{k]}(\omega),\pi_{(k}(\omega))$ for all $\omega$ in $X$) and if $F$ is as in (a), show that $F = \pi_{(k}^{-1}(\pi_{(k}(F))$ for all $k$. (Hint: the hypothesis on $F$ is that $1_F(\omega) = 1_F(\tilde{\omega})$, whenever $\tilde{\omega}$ is obtained by changing any finitely many coordinates of $\omega$.)

(c) Show that $\mu(C \cap F) = \mu(C)\mu(F)$ for every (finite) cylinder set $C$ in $X$.

(d) Show that $\mu(C \cap F) = \mu(C)\mu(F)$ for all $C$ in $\mathcal{F}$. (Hint: the collection of $C$ for which the assertion is valid is a monotone class containing the field of cylinder sets.)

(e) Conclude that $\mu(E) = 0$ or 1, and hence that $G$ acts ergodically on $X$. (Hint: Put $C = F$ in (d) and use $\mu(E) = \mu(F)$.) □

**Lemma 4.3.8.** *With the above notation, let $\Delta(G)$ denote the multiplicative group generated by $\{p_i/p_j: 1 \leq i,j \leq N\}$; then $r(G)$ is the closure (in $[0,\infty)$) of $\Delta(G)$.*

**Proof.** Fix $k$ in $\mathbb{N}$, $\sigma \in C_n$ and for $j = 1, \ldots, N$, let $C_{j,k} = \{\omega \in X: \omega(k) = j\}$. Then $\{C_{j,k}\}_{j=1}^N$ is a (measurable) partition of $X$ and clearly

$$\frac{d\mu \circ T_{\sigma,k}}{d\mu} \equiv \frac{p_{\sigma(j)}}{p_j}$$

on $C_{j,k}$. Since every element of $G$ is a product of finitely many $T_{\sigma,k}$'s (in fact, is of the form

$$T_{\sigma_1,k_1}T_{\sigma_2,k_2} \cdots T_{\sigma_m,k_m}$$

where $1 \leq k_1 < k_2 < \cdots < k_m$), it can be deduced that if $T \in G$, then there is a partition $\{D_1, \ldots, D_\ell\}$ of $X$ into cylinder sets such that

$$\frac{d\mu \circ T}{d\mu} = \sum_{i=1}^{\ell} \lambda_i 1_{D_i},$$

where $\lambda_i \in \Delta(G)$ for each $i$; it follows immediately that $r(G) \subseteq \overline{\Delta(G)}$.

For the reverse inclusion, it clearly suffices to show that $p_i/p_j \in r(G)$ for $1 \leq i,j \leq N$. So fix $i,j \leq N$ and suppose $\epsilon > 0$, $E \in \mathcal{F}$ and $\mu(E) > 0$. Choose $\delta$ so that

$$0 < \delta < \frac{p_i p_j}{p_i + p_j + p_i p_j}\mu(E).$$

Since $\mu$ is a product-measure, there exists a (finite) cylinder set, say $C$, such that $\mu(C \Delta E) < \delta$. Notice that $\delta < \mu(E)$, so that $\mu(C \cap E) > 0$. Let $E_0 = C \cap E$; thus, $\mu(E_0) > 0$.

Suppose the cylinder set $C$ is determined by the first $m$ coordinates; fix $k > m$, let $D = \{\omega \in X: \ \omega(k) = j\}$, pick $\sigma \in C_N$ so that $\sigma(j) = i$ and write $T = T_{\sigma,k} \in G$. The following assertions are clear:

(a) $\mu(C \cap D) = \mu(C)\mu(D) = p_j\mu(C)$;

(b) $\dfrac{d\mu oT}{d\mu} \equiv \dfrac{p_i}{p_j}$ on $D$; and

(c) $T(C) = C$.

Observe now that

$$\begin{aligned}\mu(E_0 \cap D) &\geq \mu(C \cap D) - \mu(C\backslash E_0)\\ &> p_j\mu(C) - \delta\\ &> p_j(\mu(E) - \delta) - \delta;\end{aligned}$$

hence, by (b) above,

$$\mu(T(E_0 \cap D)) > \frac{p_i}{p_j}[p_j(\mu(E) - \delta) - \delta];$$

$$\begin{aligned}&\mu(T(E_0 \cap D)) + \mu(E_0)\\ &\quad > \frac{p_i}{p_j}[p_j(\mu(E) - \delta) - \delta] + \mu(C) - \delta\\ &\quad = \mu(C) + \frac{1}{p_j}[p_i p_j \mu(E) - \delta(p_i p_j + p_i + p_j)]\\ &\quad > \mu(C).\end{aligned}$$

Since the relations $T(C) = C$ and $E_0 \subseteq C$ force $T(E_0 \cap D) \subseteq C$, the above inequality ensures that $\mu(T(E_0 \cap D) \cap E_0) > 0$. Set $F = (E_0 \cap D) \cap T^{-1}(E_0)$ and note that $F \subseteq E$, $T(F) \subseteq E$, $\mu(F) > 0$ and $d\mu oT/d\mu \equiv p_i/p_j$ on $F$. Since $E$ (and $\epsilon$) was (were) arbitrary, this establishes that $p_i/p_j \in r(G)$ as desired. □

**Example 4.3.9.** ($\lambda$) Let $0 < \lambda < 1$, $X_0 = \{1,2\}$, $p_1 = 1/1+\lambda$ and $p_2 = \lambda/1+\lambda$. Then

$$\left\{\frac{p_1}{p_2}, \frac{p_2}{p_1}\right\} = \{\lambda^{-1}, \lambda\}.$$

So, with $G$ as in Lemma 4.3.8 (and $n = 2$), we have $r(G) = \{0\} \cup \{\lambda^n: n \in \mathbb{Z}\}$.

(b) With the notation of Lemma 4.3.8, let $N = 3$, $\lambda_1, \lambda_2 > 0$ and

$$p_1 = \frac{1}{1 + \lambda_1 + \lambda_2}, \quad p_2 = \frac{\lambda_1}{1 + \lambda_1 + \lambda_2}, \quad p_3 = \frac{\lambda_2}{1 + \lambda_1 + \lambda_2};$$

then $\Delta(G)$ is the multiplicative group generated by $\lambda_1$ and $\lambda_2$; so, if $\log \lambda_1$ and $\log \lambda_2$ are not rational multiples of one another -- for instance, $\lambda_1 = e$, $\lambda_2 = e^{\sqrt{2}}$ -- then $r(G) = [0,\infty)$. □

**Example 4.3.10.** Let $X_0 = \{-1,1\}$ and let $X = X_0^{\mathbf{N}}$. We shall show that there exists a sequence $\{\mu_n\}_{n=1}^{\infty}$ of probability measure on $X_0$, such that if $\mu = \otimes_{n=1}^{\infty}\mu_n$ (defined on the product $\sigma$-algebra $\mathcal{F}$ on $X$), and if $G$ is as before (in this case, $G$ is the group generated by $\{T_n\}_{n=1}^{\infty}$, where

$$(T_n\omega)(k) = (-1)^{\delta_{nk}}\omega(k) \qquad )$$

then $r(G) = \{0,1\}$.

Let $\{k_j\}_{j=1}^{\infty}$ be a sequence of positive integers satisfying the following conditions:

$$\left.\begin{array}{ll} \text{(i)} & k_j < k_{j+1} \\ \text{(ii)} & k_j | k_{j+1} \text{ (i.e., } k_j \text{ divides } k_{j+1}) \\ \text{(iii)} & k_1 2^{k_1} + \cdots + k_j 2^{k_j} < k_{j+1} \end{array}\right\} \text{ for all } j \qquad (*)$$

Next, define a sequence $\{m_n\}_{n=1}^{\infty}$ of integers thus:

$$m_n = k_j \quad \text{if} \quad \sum_{i=1}^{j-1} 2^{k_i} < n \leqslant \sum_{i=1}^{j} 2^{k_i};$$

in other words, the first $2^{k_1}$ $m_n$'s are all equal to $k_1$, the next $2^{k_2}$ $m_n$'s are all equal to $k_2$, etc.

Let $\mu_n$ be the probability measure on $\{-1,1\}$ defined by

$$\mu_n(\{-1\}) = \frac{2^{m_n}}{1 + 2^{m_n}}, \quad \mu_n(\{1\}) = \frac{1}{1 + 2^{m_n}},$$

and let $\mu$ be the product measure $\mu = \otimes_{n=1}^{\infty}\mu_n$ on $X$.

**Step 1.** $\mu(\{\omega\}) = 0$ for all $\omega$ in $\Omega$.

**Proof.**

$$\mu(\{\omega\}) = \prod_{n=1}^{\infty} \mu_n(\{\omega(n)\}) \leqslant \prod_{n=1}^{\infty} \left(\frac{2^{m_n}}{1 + 2^{m_n}}\right) = \frac{1}{\prod_{n=1}^{\infty}(1 + 2^{-m_n})};$$

since

$$\sum_{n=1}^{\infty} 2^{-m_n} = 2^{k_1}\cdot 2^{-k_1} + 2^{k_2}\cdot 2^{-k_2} + \cdots = \infty,$$

conclude that $\mu(\{\omega\}) = 0$.

**Step 2.** For $j = 1,2, \ldots$, let $I_j = \{n \in \mathbb{N}: m_n = k_j\}$, $A_j = \{\omega \in X: \omega(n) = -1$ for all $n$ in $I_j\}$, and $B_j = A_j^c = \{\omega \in X: \exists n$ in $I_j$ such that $\omega(n) = +1\}$. If $\mu(E) > 0$ and if $j_0$ is any integer, there exists an integer $j > j_0$ such that $\mu(E \cap A_j) > 0$ and $\mu(E \cap B_j) > 0$.

**Proof.** Since $k_j \nearrow \infty$ as $j \to \infty$, conclude that

$$\mu(A_j) = \left[\frac{2^{k_j}}{1 + 2^{k_j}}\right]^{2^{k_j}} = \frac{1}{(1 + 2^{-k_j})^{2^{k_j}}} \to \frac{1}{e},$$

and hence $\mu(B_j) \to 1 - 1/e$. Pick a number $\gamma$ such that $(1/e <)\ 1 - 1/e < \gamma < 1$; next pick an integer $j_1 > j_0$ such that if $j \geq j_1$ and $C_j = A_j$ or $B_j$, then $\mu(C_j) < \gamma$; then pick an integer $j_2 > j_1$ such that

$$\gamma^{j_2 - j_1 + 1} < \mu(E).$$

If $\mu(E \cap A_j) = 0$ or $\mu(E \cap B_j) = 0$ for each $j$ satisfying $j_1 \leq j \leq j_2$, let $C_j = B_j$ or $A_j$ and note that

$$E \subseteq \bigcap_{j=j_1}^{j_2} C_j \ (\text{mod } \mu).$$

Notice now that for distinct $j$'s, the $C_j$'s are cylinder sets determined by disjoint sets of coordinates -- i.e., $\{C_j: j_1 \leq j \leq j_2\}$ are "independent events" -- and hence

$$\mu(E) \leq \mu\left[\bigcap_{j_1 \leq j \leq j_2} C_j\right] = \prod_{j=j_1}^{j_2} \mu(C_j) < \gamma^{j_2 - j_1 + 1} < \mu(E);$$

this contradiction completes the proof.

**Step 3.** Let $\ell_1, \ldots, \ell_n$ be integers satisfying $|\ell_j| \leq 2^{k_j}$ for $1 \leq j \leq n$; then $\Sigma_{j=1}^n \ell_j k_j = 0$ implies $\ell_j = 0$ for $1 \leq j \leq n$.

**Proof.** If not, assume without loss of generality that $\ell_n \neq 0$, and observe that

$$k_n \leq |\ell_n k_n| = \left|\sum_{j<n} \ell_j k_j\right| \leq \sum_{j<n} 2^{k_j} k_j,$$

which contradicts the choice of the $k_j$'s (cf. (iii) (*)).

**Step 4.** $r(\mathcal{G}) = \{0,1\}$.

**Proof.** Temporarily fix an arbitrary integer $j > 0$, and let $E = \{\omega \in X: \omega(n) = 1$ for $1 \leqslant n \leqslant 2^{k_1} + \cdots + 2^{k_{j-1}}\}$. Clearly $E \in \mathcal{F}$ and $\mu(E) > 0$. Suppose there exists $F$ in $\mathcal{F}$ and $T$ in $G$ such that $\mu(F) > 0$ and $F \cup TF \subseteq E$; then it must be the case that $T$ is in the group generated by $\{T_n: m_n \geqslant k_j\}$; it follows from the assumption that $k_i | k_{i+1}$ that

$$\frac{d\mu \circ T}{d\mu} = \sum_{i=1}^{\ell} \lambda_i 1_{D_i},$$

where $\{D_i\}_{i=1}^{\ell}$ is a partition of $X$ (into cylinder sets defined by coordinates after $2^{k_1} + \cdots + 2^{k_{j-1}}$) and each $\lambda_i$ is in the cyclic (multiplicative) group generated by $2^{k_j}$. Since $j$ was arbitrary, conclude that $r(G) \subseteq \{0,1\}$.

To prove that $0 \in r(G)$, it suffices to show that if $E \in \mathcal{F}$, $\mu(E) > 0$ and $\epsilon > 0$, there exist $F \in \mathcal{F}$ and $T \in G$ such that $F \cup TF \subseteq E$, $\mu(F) > 0$ and $(d\mu o T/d\mu)(\omega) \notin [\epsilon, \epsilon^{-1}]$ for all $\omega$ in $F$ (where we have assumed $\epsilon < 1$, as we clearly may). (Then, either

$$\left\{F \cap \left\{\omega : \frac{d\mu o T}{d\mu}(\omega) < \epsilon\right\}, T\right\}$$

or

$$\left\{T\left[F \cap \left\{\omega : \frac{d\mu o T}{d\mu}(\omega) > \epsilon\right\}\right], T^{-1}\right\}$$

will do the job.)

First fix $j_0$ such that $2^{-k_j} < \epsilon$ for $j \geqslant j_0$. Let

$$n_0 = \sum_{i=1}^{j_0 - 1} 2^{k_i};$$

since $\mu(E) > 0$, there exists $\omega_1^0, \ldots, \omega_{n_0}^0$ in $\{-1,1\}$ such that if $C = \{\omega \in \Omega: \omega(n) = \omega_n^0$ for $1 \leqslant n \leqslant n_0\}$, then $\mu(E \cap C) > 0$. By Step 2 applied to $E \cap C$ and this $j_0$, there exists an integer $\tilde{j} > j_0$ such that $\mu(E \cap C \cap A_{\tilde{j}}) > 0$ and $\mu(E \cap C \cap B_{\tilde{j}}) > 0$. Since $G$ acts ergodically on $X$, (cf. Ex. (4.3.11)) there exists $T$ in $G$ such that

$$\mu(T(E \cap C \cap A_{\tilde{j}}) \cap (E \cap C \cap B_{\tilde{j}})) > 0.$$

Let $F = (E \cap C \cap A_{\tilde{j}}) \cap T^{-1}(E \cap C \cap B_{\tilde{j}})$; then $F \in \mathcal{F}$, $\mu(F) > 0$ (since $\mu(TF) > 0$) and $F \cup TF \subseteq E$. It follows from the definition of $C$ that $T$ must belong to the group generated by $\{T_n: n > n_0\}$, and also $T \neq \epsilon$, since $A_{\tilde{j}} \cap B_{\tilde{j}} = \emptyset$. So, there exist integers $n_1 < n_2 < \cdots < n_p$ such that $n_0 < n_1$ and $T = T_{n_1} T_{n_2} \cdots T_{n_p}$. Argue (as in the proof of $r(G) \subseteq \{0,1\}$) that the range of $d\mu o T/d\mu$ is contained in the cyclic group generated by $2^{k_{j_0}}$, which in turn, is contained in $(0,\epsilon) \cup \{1\} \cup$

$(\epsilon^{-1},\infty)$, since $2^{-k_{j_0}} < \epsilon$. Hence, to complete the proof, it suffices to show that $(d\mu o T/d\mu)(\omega) \neq 1$ for all $\omega$ in $F$.

Choose $N$ such that $\{m_{n_1}, \ldots, m_{n_p}\} \subseteq \{k_1, \ldots, k_N\}$; for $\omega$ in $X$ and $1 \leq j \leq N$, define $\ell_j(\omega) = \Sigma\{\omega(n_i): m_{n_i} = k_j, 1 \leq i \leq p\}$. Since for any $n$,

$$\frac{d\mu \circ T_n}{d\mu}(\omega) = 2^{\omega(n)m_n},$$

conclude that

$$\frac{d\mu \circ T}{d\mu}(\omega) = 2^{\Sigma_{j=1}^N \ell_j(\omega)k_j}.$$

However, since $A_{\tilde{j}} \cap B_{\tilde{j}} = \emptyset$, at least one $n_i$ belongs to $I_{\tilde{j}}$, and so (by the definition of $A_{\tilde{j}}$), $\ell_{\tilde{j}}(\omega) \leq -1$ for all $\omega$ in $A_{\tilde{j}}$; since clearly $|\ell_j(\omega)| \leq 2^{k_j}$ for $1 \leq j \leq N$, conclude from Step 3 that $(d\mu o T/d\mu)(\omega) \neq 1$ for all $\omega$ in $A_{\tilde{j}}$, and, in particular, for all $\omega$ in $F$, and the proof is complete. □

## Exercises

**(4.3.11)** Suppose $G$ is a countable group of automorphisms of a general measure space $(X,\mathcal{F},\mu)$, and suppose $G$ acts ergodically. If $E,F \in \mathcal{F}$ and $\mu(E) > 0$, $\mu(F) > 0$, show that there exists a $T$ in $G$ such that $\mu(TE \cap F) > 0$. (Hint: the set $\tilde{E} = \cup_{T \in G} TE$ has positive measure and $T\tilde{E} = \tilde{E}$ for all $T$ in $G$.) □

We shall conclude this section with a very elegant result of von Neumann which establishes an almost complete lexicon between a free and ergodic action $t \to T_t$ (of a countable group as automorphisms of a measure space $(X,\mathcal{F},\mu)$) and the type of the factor $M = L^\infty(X,\mathcal{F},\mu) \otimes_\alpha G$ (where $\alpha_t(f) = f \circ T_t^{-1}$). Before getting to that theorem, let us digress with a preliminary technical lemma.

The following observations will be needed in the proof of the lemma: recall that the Schauder fixed-point theorem asserts that any continuous self-map of a compact convex subset $K$ of a locally convex topological vector space has a fixed point. An easy generalization is the following: if $K$ is as above and if $\{T_i: i \in I\}$ is any family of pairwise commuting continuous affine self-maps of $K$ (affine means that $T_i$ respects convex combinations), then $K$ contains a point which is fixed by each $T_i$. (Reason: let $K_i = \{x \in K: T_i x = x\}$; the hypothesis on the $T_j$'s and the Schauder fixed point theorem ensure that $K_i$ is a compact convex non-empty set which is mapped into itself by each $T_j$; repeated application of the Schauder fixed point theorem shows that the collection $\{K_i: i \in I\}$ of compact subsets of $K$ has the finite-intersection property, and hence $\cap K_i \neq \phi$, as asserted.)

**Lemma 4.3.12.** *Let $M$ be a semifinite von Neumann algebra with a* fns *trace $\tau$. Let $A$ be a maximal abelian von Neumann subalgebra of $M$. If there exists a normal norm-one projection $E$ of $M$ onto $A$, then $\tau|A_+$ is semifinite and $\tau \circ E = \tau$.*

**Proof.** Fix $x$ in $M_+$ and let $C(x)$ be the convex hull of $\{uxu^*: u \in \mathcal{U}(A)\}$. Since $C(x)$ is norm-bounded, it follows that the $\sigma$-weak closure $K(x)$ of $C(x)$ is a $\sigma$-weakly compact convex subset of $M$. The abelian group $\mathcal{U}(A)$ clearly acts (via inner conjugation: $T_u(y) = uyu^*$, $y \in K(x)$) as a family of pairwise commuting $\sigma$-weakly continuous affine self-maps of $K(x)$. So, by the discussion preceding the lemma, there is a point $x_0$ in $K(x)$ such that $ux_0u^* = x_0$ for all $u$ in $\mathcal{U}(A)$. Since $A$ is maximal abelian, it follows that $x_0 \in A$. (Note that $x \in M_+ \Rightarrow K(x) \subseteq M_+ \Rightarrow x_0 = x_0^*$.) In particular, $K(x) \cap A \neq \phi$.

Suppose now that $F$ is any normal norm-one projection of $M$ onto $A$. Then, by Prop. 2.6.4, notice that for any $u$ in $\mathcal{U}(A)$, $F(uxu^*) = uF(x)u^* = F(x)$, since $A$ is abelian; thus $F$ is constant on $C(x)$. Since the normality of $F$ implies that $F$ is $\sigma$-weakly continuous, conclude that $F$ is constant on $K(x)$. Since $F$ fixes points in $A$, conclude that $K(x) \cap A = \{F(x)\}$. Since $x \in M_+$ was arbitrary, conclude that any two normal norm-one projections of $M$ onto $A$ must be equal.

Suppose we can show that $\tau|A_+$ is semifinite. Since $\tau$ is a trace, it would then follow (cf. Remark 2.6.9 (a)) that there exists a normal norm-one projection $F$ of $M$ onto $A$ such that $\tau \circ F = \tau$. The uniqueness statement of the last paragraph would then show that $E = F$, and so $\tau \circ E = \tau$. Thus it suffices to establish the semifiniteness of $\tau|A_+$.

Suppose $x \in \mathfrak{D}_\tau$ -- i.e., $x \in M_+$ and $\tau(x) < \infty$. The traciality of $\tau$ implies that $\tau(y) = \tau(x)$ for all $y$ in $C(x)$. Since $\tau$ is $\sigma$-weakly lower-semicontinuous (cf. Prop. 2.4.9), infer that $\tau(y) \leqslant \tau(x)$ for all $y$ in $K(x)$; in particular, $\tau(Ex) < \infty$, since $Ex \in K(x)$ (we have shown, in fact, that $K(x) \cap A = \{Ex\}$).

Since $\tau$ is semifinite, there is an increasing net $\{x_i\}$ in $\mathfrak{D}_\tau$ such that $x_i \nearrow 1$; since $E$ is normal, $Ex_i \nearrow 1$; however, by the last paragraph, $Ex_i \in D_{\tau|A_+}$; this shows that $\tau|A_+$ is semifinite. □

Suppose now that $t \to T_t$ is a free and ergodic action of a countable group $G$ as automorphisms of a separable and $\sigma$-finite measure space $(X,\mathfrak{F},\mu)$. Let $t \to \alpha_t$ be the induced action of $G$ on $M = L^\infty(X,\mathfrak{F},\mu)$. Then, the crossed-product $\tilde{M} = M \otimes_\alpha G$ is a factor. The type of $\tilde{M}$ -- in the Murray-von Neumann classification -- is determined as below.

**Theorem 4.3.13.** (a) *$\tilde{M}$ is of type* III *if and only if there does not exist a $\sigma$-finite positive measure $\nu$ which is equivalent to $\mu$ (in the sense of mutual absolute continuity) such that $\nu \circ T_t = \nu$ for all $t$ in $G$.*

(b) *Suppose there exists a $G$-invariant $\sigma$-finite positive measure $\nu$ which is equivalent to $\mu$ (so that $\tilde{M}$ is semifinite, by* (a)*). Then,*

(i) $\tilde{M}$ *is of type* I *if and only if the measure space* $(X,\mathcal{F},\mu)$ *contains atoms*;

(ii) $\tilde{M}$ *is of type* II *if and only if the measure space* $(X,\mathcal{F},\mu)$ *contains no atoms*;

(iii) $\tilde{M}$ *is a factor of finite type if and only if* $\nu$ *is a finite measure.*

**Proof.** (a) Suppose such a measure $\nu$ exists. Let $\tau$ be the fns trace on $M$ given by $\tau(f) = \int f d\nu$. Conclude from Ex. (4.1.6) (b) and Ex. (4.2.11) (a) that the equation $\tilde{\tau}(\tilde{x}) = \tau(\tilde{x}(\epsilon,\epsilon))$ defines a fns trace on $\tilde{M}$, whence $\tilde{M}$ is semifinite.

Conversely, suppose $\tilde{M}$ is semifinite, and $\tilde{\tau}$ is some fns trace on $\tilde{M}$. Since the action $\alpha$ is free, Corollary 4.1.9 ensures that $\pi_\alpha(M)$ is a maximal abelian von Neumann subalgebra of $\tilde{M}$; hence, it follows from Ex. 4.1.6 and Lemma 4.3.12 that $\tau = \tilde{\tau}|\pi_\alpha(M)$ is a fns trace on $\pi_\alpha(M)$. Use the isomorphism $\pi_\alpha$ to transfer $\tau$ to a fns trace on $M$, which we shall continue to call $\tau$. It follows (cf. the opening remarks of Section 2.6) that there is a $\sigma$-finite positive measure $\nu$, which is equivalent to $\mu$ (since $\tau$ is faithful) such that $\tau(f) = \int fd\nu$ for $f$ in $M_+$. Observe, then, that if $f \in M_+$ and $t \in G$,

$$\begin{aligned}
\int f\, d\nu = \tau(f) &= \tilde{\tau}(\pi(f)) \\
&= \tilde{\tau}(\lambda(t)\pi(f)\lambda(t)^*) \quad \text{(since } \tilde{\tau} \text{ is a trace)} \\
&= \tilde{\tau}(\pi(\alpha_t(f))) \\
&= \tau(f \circ T_t^{-1}) \\
&= \int (f \circ T_t^{-1}) d\nu \\
&= \int f\, d(\nu \circ T_t);
\end{aligned}$$

setting $f = 1_E$ for arbitrary $E$ in $\mathcal{F}$, conclude that $\nu = \nu \circ T_t$.

(b) Suppose $(X,\mathcal{F},\mu)$ contains an atom $E$ -- i.e., $E \in \mathcal{F}$, $\mu(E) > 0$ and whenever $F \in \mathcal{F}$ and $F \subseteq E$, either $\mu(F) = 0$ or $\mu(E\backslash F) = 0$. Clearly then $1_E$ is a minimal projection in $M$. Let $e = \pi(1_E)$. Since $e \neq 0$, it will follow that $\tilde{M}$ is of type I if we show that $e$ is a minimal projection in $\tilde{M}$. Suppose $f \in \mathcal{P}(\tilde{M})$ and $0 \neq f \leqslant e$. Since $e$ is minimal in $\pi(M)$, it follows that $x \in \pi(M) \Rightarrow exe = \lambda e$ for some $\lambda$ in $\mathbb{C}$ (cf. Ex. (4.3.1) (a), or, at least the hint for that exercise); hence, $x \in \pi(M) \Rightarrow fx = fex = fexe = f(\lambda e) = \lambda fe = \lambda f$ and similarly $xf = \lambda f$; in particular $f \in \pi(M)' \cap \tilde{M} = \pi(M)$ (by Cor. 4.1.9), and as $e$ is minimal in $\pi(M)$, conclude that $f = e$.

If $(X,\mathcal{F},\mu)$ is non-atomic, clearly so is $(X,\mathcal{F},\nu)$. So, there exist sets $E_n \in \mathcal{F}$ such that $\nu(E_n) > 0$ for all $n$ and $\nu(E_n) \to 0$. If $\tau(f) = \int fd\nu$ for $f$ in $M_+$ and $\tilde{\tau}$ is the induced fns trace on $\tilde{M}$ (see proof of (a)), then $\tilde{\tau}|\mathcal{P}(\tilde{M})$ is a dimension function for $\tilde{M}$ and $\tilde{\tau}(\pi(1_{E_n})) \to 0$; thus $\tilde{M}$ is of of type II.

Since the possibilities "$(X,\mathcal{F},\mu)$ has atoms" and "$(X,\mathcal{F},\mu)$ has no atoms" are mutually exclusive, as are the possibilities "$M$ is of type I" and "$M$ is of type II", the previous two paragraphs establish the validity of (i) and (ii). For (iii), with $\tau$ as above, note that $\tau(1) = \nu(X)$. □

**Example 4.3.14.** Let $\tilde{G}$ be a second countable locally compact group, with a left Haar measure $\mu$ (defined on the Borel $\sigma$-algebra $\mathcal{F}$ of $\tilde{G}$). Then $(\tilde{G},\mathcal{F},\mu)$ is separable and $\sigma$-finite. Suppose $G$ is a countable dense subgroup of $\tilde{G}$. Then $G$ acts on $\tilde{G}$ as left translations: $T_t\tilde{g} = t\tilde{g}$ for $t$ in $G$ and $\tilde{g}$ in $\tilde{G}$. Then it is easily seen that $t \to T_t$ is a free, ergodic action of $G$ on $(\tilde{G},\mathcal{F},\mu)$ as measure-preserving automorphisms. If $t \to \alpha_t$ is the induced action of $G$ on $M = L^\infty(\tilde{G},\mathcal{F},\mu)$ it follows that $\tilde{M} = M \otimes_\alpha G$ is a semifinite factor. Then $\tilde{M}$ is of (i) type I, (ii)$_1$ type II$_1$, or (ii)$_\infty$ type II$_\infty$ if and only if (i) $\tilde{G}$ is discrete and $G = \tilde{G}$, (ii)$_1$ $\tilde{G}$ is not discrete, but compact, or (ii)$_\infty$ $\tilde{G}$ is not discrete and non-compact (since $\mu$ is finite if and only if $\tilde{G}$ is compact). Examples are given by (i)$_n$ $G = \tilde{G} = \mathbb{Z}_n$, the cyclic group of order $n$; (i)$_\infty$ $G = \tilde{G} = \mathbb{Z}$, the infinite cyclic group; (ii)$_1$ $\tilde{G} = T = \{z \in \mathbb{C}: |z| = 1\}$ (under multiplication) and $G = \{e^{in\theta}: n \in \mathbb{Z}\}$ where $\theta/2\pi$ is irrational; and (ii)$_\infty$ $\tilde{G} = \mathbb{R}$ and $G = Q$ (the rational numbers). □

**Example 4.3.15.** In order to use Theorem 4.3.13 to construct examples of factors of type III, one must have some condition which will ensure the non-existence of an equivalent invariant measure. One such is given by:

**Assertion:** If $\mathcal{G}$ is an ergodic group of automorphisms of $(X,\mathcal{F},\mu)$, if $\mathcal{G}_0 = \{T \in \mathcal{G} : \mu \circ T = \mu\} \neq \mathcal{G}$, and if $\mathcal{G}_0$ also acts ergodically on $(X,\mathcal{F},\mu)$, then there exists no $\sigma$-finite positive measure $\nu$ which is equivalent to $\mu$ and $\mathcal{G}$-invariant.

**Proof.** Suppose such a measure $\nu$ exists. If $g = d\nu/d\mu$, $T \in \mathcal{G}_0$ and $f$ is any non-negative (measurable) function on $X$, then

$$\int (f \circ T)(g \circ T)d\mu = \int fg\, d(\mu \circ T^{-1}) = \int fg\, d\mu = \int f\, d\nu$$

$$= \int fd(\nu \circ T^{-1}) = \int (f \circ T)d\nu = \int (f \circ T)g\, d\mu;$$

since $f$ is arbitrary and $T$ is an automorphism, conclude that $g \circ T = g$ a.e. $(\mu)$. Since $T$ was arbitrary and $\mathcal{G}_0$ acts ergodically on $(X,\mathcal{F},\mu)$ conclude (by observing that $g^{-1}(E)$ has full or zero measure for every Borel set $E$ in $\mathbb{R}$) that $g$ is constant: $g = r > 0$ (say). Then $\mu = r^{-1}\nu$ is $\mathcal{G}$-invariant, contradicting the assumption that $\mathcal{G}_0 \neq \mathcal{G}$. This completes the proof of the assertion.

Let $(X,\mathcal{F},\mu)$ be the real line with Lebesgue measure; let $\mathcal{G}$ be the group of automorphisms $\{T_{p,q}: p,q \in Q, p > 0\}$ where $T_{p,q}(t) = pt + q$ for $t$ in $\mathbb{R}$. Clearly $T_{1,0}$ is the identity element for $\mathcal{G}$ and it is easily seen that $T_{p,q}$ is free if $(p,q) \neq (1,0)$. In the notation of the assertion, we have $G_0 = \{T_{1,q}: q \in Q\}$, which clearly acts ergodically on $\mathbb{R}$ (cf. Example 4.3.14 (ii)$_\infty$). It follows that $L^\infty(\mathbb{R}) \otimes_\alpha G$ is a factor of type III, where $\alpha_T(f) = f \circ T^{-1}$. □

**Exercises**

**(4.3.16)** Let $G = \{T_{p,q}: p,q \in Q, p > 0\}$ act on $L^\infty(\mathbb{R})$ as in Example 4.3.15. Show that $L^\infty(\mathbb{R}) \otimes_\alpha G$ is of type $\mathrm{III}_1$. (Hint: it is enough to show that $p \in r(G)$ for all $p \in Q \cap (0,\infty)$; if $\mu(E) > 0$, also $\mu(p^{-1}E) > 0$, where $p^{-1}E = \{p^{-1}t: t \in E\}$; since $\{T_{1,q}: q \in Q\}$ is ergodic, pick $q \in Q$ such that $\mu((p^{-1}E + q) \cap E) > 0$; let $F = (p^{-1}E + q) \cap E$ and observe that $F \cup TF \subseteq E$ where $T = T_{p,-qp}$; and $d\mu o T/d\mu \equiv p$.)

**(4.3.17)** Let $G = \{T_{\lambda^n,q} : n \in \mathbb{Z}, q \in Q\}$ where $0 < \lambda < 1$, and let $\alpha$ be the natural action of $G$ on $L^\infty(\mathbb{R})$. Show that $L^\infty(\mathbb{R}) \otimes_\alpha G$ is of type $\mathrm{III}_\lambda$. (Hint: argue as in Ex. (4.3.16) to see that $\{\lambda^n: n \in \mathbb{Z}\} \subseteq r(G)$; since

$$\bigcup_{T \in G} \operatorname{ran} \frac{d\mu \circ T}{d\mu} \subseteq \{\lambda^n : n \in \mathbb{Z}\},$$

infer that $r(G) \subseteq \{0\} \cup \{\lambda^n: n \in \mathbb{Z}\}$. □

Notice that while Theorem 4.3.13 is completely satisfactory as far as the Murray-von Neumann classification of factors is concerned, it does not say anything concerning Connes' subclassification of type III factors. To do this, we must augment that theorem by the following result, whose proof we will not spell out, since it is an easy consequence of Proposition 4.2.9 and the discussion up to Definition 4.2.10.

**Proposition 4.3.18.** *Let $G$ be a countable group of automorphisms of a separable and σ-finite measure space $(X,\mathfrak{F},\mu)$; assume that $G$ acts freely and ergodically. Let α denote the induced action of $G$ on $L^\infty(X,\mathfrak{F},\mu)$ and let $M$ be the factor $L^\infty(X,\mathfrak{F},\mu) \otimes_\alpha G$.*

(a) *$\breve{M}$ is of type* $\mathrm{III}_0 \Leftrightarrow r(G) = \{0,1\}$;

(b) *$\breve{M}$ is of type* $\mathrm{III}_\lambda \Leftrightarrow r(G) = \{0\} \cup \{\lambda^n: n \in \mathbb{Z}\}$; $(0 < \lambda < 1)$

(c) *$\breve{M}$ is of type* $\mathrm{III}_1 \Leftrightarrow r(G) = [0,\infty)$. □

**Corollary 4.3.19.** *Let $G$ be as in Prop.* 4.3.18. *The following conditions are equivalent:*

(i) *there exists a σ-finite measure ν which is equivalent to μ and satisfies $\nu \circ T = \nu$ for all $T$ in $G$;*
(ii) $r(G) = \{1\}$.

**Proof.** This is an immediate consequence of the equality $S(\tilde{M}) = r(G\hat{\ })$, Theorem 4.3.13 (a) and Proposition 3.4.6. □

## 4.4. Continuous Crossed-Products and Takesaki's Duality Theorem

The symbol $G$ will, throughout this section, denote a locally compact (Hausdorff) group which, for convenience, will be assumed to satisfy the second axiom of countability: i.e., $G$ has a countable base of open sets. It follows from Urysohn's metrization theorem (second countable and regular implies metrizable) that $G$ is metrizable, separable and $\sigma$-compact. It follows that $L^2(G)$ -- with respect to a fixed left Haar measure, denoted simply by $ds$ -- is separable.

If $\mathcal{H}$ is any separable Hilbert space, the symbol $\tilde{\mathcal{H}}$ will denote the Hilbert space $L^2(G;\mathcal{H})$, a typical member of which is a weakly measurable function $\tilde{\xi}$: $G \to \mathcal{H}$ satisfying $\int \|\tilde{\xi}(s)\|^2 ds < \infty$. (As is customary, two functions will be identified if they agree almost everywhere.) A useful fact -- which will help overcome annoying measurability questions -- is that the space $C_c(G;\mathcal{H})$, of strongly continuous functions $\xi$: $G \to \mathcal{H}$ which are compactly supported, is dense in $\tilde{\mathcal{H}}$. It is a standard fact that $\tilde{\mathcal{H}}$ is canonically isomorphic to $\mathcal{H} \otimes L^2(G)$, whereby $\xi \otimes f$ corresponds to the function $s \to f(s)\xi$; when convenient, we shall employ this description of $\tilde{\mathcal{H}}$.

As usual, we shall let $t \to \lambda_t$ denote the left-regular representation of $G$ in $L^2(G)$: $(\lambda_t f)(s) = f(t^{-1}s)$; we shall let $t \to \lambda(t)$ denote the (clearly strongly continuous unitary) representation of $G$ in $\tilde{\mathcal{H}}$ given by $(\lambda(t)\tilde{\xi})(s) = \tilde{\xi}(t^{-1}s)$ -- or, in the picture $\tilde{\mathcal{H}} = \mathcal{H} \otimes L^2(G)$, $\lambda(t) = 1 \otimes \lambda_t$.

Suppose now that $(M,G,\alpha)$ is a dynamical system -- in the sense of Definition 3.2.1 (b) -- where $M$ is a von Neumann algebra of operators on $\mathcal{H}$. It is fairly easily established that there is a normal *-isomorphism $\pi_\alpha$: $M \to \mathcal{L}(\tilde{\mathcal{H}})$ such that

$$(\pi_\alpha(x)\tilde{\xi})(s) = \alpha_{s^{-1}}(x)\tilde{\xi}(s).$$

(First show that $\pi_\alpha(x)$ maps $C_c(G;\mathcal{H})$ into itself and that $\|\pi_\alpha(x)\tilde{\xi}\| \leq \|x\|\|\xi\|$ for $\xi$ in $C_c(G;\mathcal{H})$; the rest of the verification is routine.) As in the discrete case, when there is no possibility of confusion, we will sometimes drop the qualifying subscript $\alpha$. Also, as in the discrete case, we have the following basic commutation relations, which are easily verified:

(4.4.1) if $t \in G$ and $x \in M$, $\pi(\alpha_t(x)) = \lambda(t)\pi(x)\lambda(t)^*$.

**Definition 4.4.1.** (a) Two dynamical systems $(M_i,G,\alpha^i)$, $i = 1,2$, are said to be isomorphic if there exists a von Neumann algebra isomorphism $\pi$: $M_1 \to M_2$ such that $\pi \circ \alpha^1_t = \alpha^2_t \circ \pi$ for all $t$ in $G$.

(b) A dynamical system $(M,G,\alpha)$, where $M \subseteq \mathcal{L}(\mathcal{H})$, is said to be unitarily implemented if there exists a strongly continuous unitary representation $t \to u_t$ of $G$ in $\mathcal{H}$ such that $u_t M u_t^* = M$ for all $t$, and $\alpha_t(x) = u_t x u_t^*$ for all $x$ in $M$ and $t$ in $\mathbb{R}$ □

**Example 4.4.2.** (a) If $\phi$ is a fns weight on $M$, and if $\beta$ is the action of $\mathbb{R}$ on $\pi_\phi(M)$ given by $\beta_t(x) = \Delta_\phi^{it} x \Delta_\phi^{-it}$ for $x$ in $\pi_\phi(M)$ and $t$ in $\mathbb{R}$, then $(\pi_\phi(M),\mathbb{R},\beta)$ is unitarily implemented, and $\pi_\phi$ establishes an isomorphism of the dynamical systems $(M,\mathbb{R},\sigma^\phi)$ and $(\pi_\phi(M),\mathbb{R},\beta)$.

(b) If $(M,G,\alpha)$ is a dynamical system, then since $\pi_\alpha: M \to \mathfrak{L}(\tilde{\mathcal{H}})$ is a normal *-isomorphism of $M$ into $\mathfrak{L}(\tilde{\mathcal{H}})$, it follows that $\pi_\alpha(M)$ is a von Neumann algebra. If $\beta$ is the action of $G$ on $\pi_\alpha(M)$ defined by $\beta_t(x) = \lambda(t)x\ \lambda(t)^*$ for $x$ in $\pi_\alpha(M)$ and $t$ in $G$, the equation (4.4.1) says that (this is indeed an action of $G$ on $\pi_\alpha(M)$ and) the dynamical systems $(M,G,\alpha)$ and $(\pi_\alpha(M),G,\beta)$ are isomorphic; thus any dynamical system is isomorphic to a unitarily implemented one. □

**Definition 4.4.3.** (a) If $(M,G,\alpha)$ is a dynamical system, let $\tilde{M} = M \otimes_\alpha G$ be the von Neumann algebra of operators on $\tilde{\mathcal{H}}$ defined by $\tilde{M} = (\pi_\alpha(M) \cup \lambda(G))''$.

(b) Let $\tilde{M}_0 = \left\{ \sum_{i=1}^{n} \pi_\alpha(x_i)\lambda(t_i): \quad x_i \in M, \ t_i \in G, \ n = 1,2, \ldots \right\}$. □

It is an immediate consequence of equation (4.4.1) that $\tilde{M}_0$ is a self-adjoint subalgebra of $\tilde{M}$, and hence $\tilde{M}_0$ is $\sigma$-weakly dense in $\tilde{M}$.

**Proposition 4.4.4.** *If $\pi$: $M_1 \to M_2$ is an isomorphism of von Neumann algebras, which establishes an isomorphism of dynamical systems $(M_i,G,\alpha^i)$, $i = 1,2$, then there exists an isomorphism*

$$\tilde{\pi} : M_1 \otimes_{\alpha^1} G \to M_2 \otimes_{\alpha^2} G$$

*such that $\tilde{\pi}(\pi_{\alpha^1}(x)) = \pi_{\alpha^2}(\pi(x))$ for $x$ in $M$ and $\tilde{\pi}(\lambda_\alpha(t)) = \lambda_\beta(t)$ for $t$ in $G$.*

**Proof.** The proof is based on a theorem of Dixmier (for the proof of which the reader may consult [Dix]) which says that if $\pi$: $M_1 \to M_2$ is an isomorphism of von Neumann algebras, where $M_i \subseteq \mathfrak{L}(\mathcal{H}_i)$, then there is a Hilbert space $\mathcal{K}$ and a unitary operator $u$: $\mathcal{H}_1 \otimes \mathcal{K} \to \mathcal{H}_2 \otimes \mathcal{K}$ such that $u(x \otimes 1)u^* = \pi(x) \otimes 1$ for all $x$ in $M_1$.

Since the proposition is clearly true when the isomorphism $\pi$ is spatial (i.e., if there is a unitary operator $u$: $\mathcal{H}_1 \to \mathcal{H}_2$ such that $uxu^* = \pi(x)$ for $x$ in $M_1$), it suffices, in view of Dixmier's result, to prove the proposition in case $\mathcal{H}_2 = \mathcal{H}_1 \otimes \mathcal{K}$ and $\pi(x) = x \otimes 1$ for $x$ in $M_1$. It is clear, in this case, that

$$\tilde{\mathcal{H}}_2 = \mathcal{H}_1 \otimes \mathcal{K} \otimes L^2(G) \cong \tilde{\mathcal{H}}_1 \otimes \mathcal{K},$$

and that the map $\tilde{\pi}$: $\tilde{M}_1 \to \tilde{M}_2$ defined by $\tilde{\pi}(\tilde{x}) = \tilde{x} \otimes 1$ does the job. □

A consequence of Proposition 4.4.4 and Example 4.4.2 (b) is that while dealing with crossed-products, we may assume that the underlying dynamical system is unitarily implemented.

**Lemma 4.4.5.** *Suppose the dynamical system $(M,G,\alpha)$ is unitarily implemented by the unitary representation $t \to u_t$ of $G$ in $\mathcal{H}$. Then the equation $(w\xi)(s) = u_s\xi(s)$ defines a unitary operator $w$ on $\mathcal{H}$, satisfying*

$$w\pi_\alpha(x)w^* = x \otimes 1, \; x \in M$$

$$w\lambda(s)w^* = u_s \otimes \lambda_s, \; s \in G;$$

*in particular, $\widetilde{M}$ is spatially isomorphic to $\{x \otimes 1, u_s \otimes \lambda_s: x \in M, s \in G\}''$.*

**Proof.** The verification is routine, and left as an exercise for the reader. □

Assume, henceforth, that $G$ is abelian, and let $\Gamma$ denote its dual group. If $\gamma \in \Gamma$, let $v_\gamma$ denote the unitary operator on $L^2(G)$ defined by $(v_\gamma\xi)(t) = \langle t,\gamma\rangle^{-1}\xi(t)$. It is clear that $\gamma \to v_\gamma$ is a strongly continuous unitary representation of $\Gamma$ in $L^2(G)$. (Recall that $\Gamma$ is topologized precisely so as to make the map $\gamma \to \langle t,\gamma\rangle$ continuous for each $t$ in $G$.) Consequently, if we set $v(\gamma) = 1 \otimes v_\gamma$, then $\gamma \to v(\gamma)$ is a strongly continuous unitary representation of $\Gamma$ in $H$.

**Lemma 4.4.6.** *With the foregoing notation,*

(a) $v(\gamma)\lambda(t)v(\gamma)^* = \langle t,\gamma\rangle^{-1}\lambda(t)$, *if* $\gamma \in \Gamma$, $t \in G$;

(b) $\{v(\gamma): \gamma \in \Gamma\} \subseteq \pi_\alpha(M)'$.

*Consequently, there is an action $\tilde{\alpha}$ of $\Gamma$ on $\widetilde{M}$ given by $\tilde{\alpha}_\gamma(\tilde{x}) = v(\gamma)\tilde{x}v(\gamma)^*$, $\tilde{x} \in \widetilde{M}$.*

**Proof.** (a) is a consequence of the readily verified equation $v_\gamma\lambda_t v_{-\gamma} = \langle t,\gamma\rangle^{-1}\lambda_t$, while (b) is a routine computation. The last statement follows from (a), (b), the definition of $\widetilde{M}$ and (the extension from $\mathbb{R}$ to general $G$ of) Ex. (2.3.4) (b). □

**Lemma 4.4.7.** *If $\pi$, $(M_i,G,\alpha^i)$, $i = 1,2$, and $\tilde{\pi}$ are as in Proposition 4.4.4, then $\tilde{\pi} \circ \tilde{\alpha}^1_\gamma = \tilde{\alpha}^2_\gamma \circ \tilde{\pi}$ for all $\gamma$ in $\Gamma$; consequently,*

$$(M_1 \otimes_{\alpha^1} G) \otimes_{\tilde{\alpha}^1} \Gamma \cong (M_2 \otimes_{\alpha^2} G) \otimes_{\tilde{\alpha}^2} \Gamma.$$

**Proof.** Since $\tilde{\pi}$ and $\tilde{\alpha}^i_\gamma$ are normal *-homomorphisms, it suffices to verify that $\tilde{\pi}(\tilde{\alpha}^1_\gamma(\tilde{x})) = \tilde{\alpha}^2_\gamma(\tilde{\pi}(\tilde{x}))$, for $\tilde{x}$ in $\pi_{\alpha^1}(M_1) \cup \lambda(G)$; this follows immediately from Prop. 4.4.4 and Lemma 4.4.6. (For instance, if $x \in M_1$,

$$\tilde{\pi}(\tilde{\alpha}^1_\gamma(\pi_{\alpha^1}(x))) = \tilde{\pi}(\pi_{\alpha^1}(x)) = \pi_{\alpha^2}(\pi(x))$$
$$= \tilde{\alpha}^2_\gamma(\pi_{\alpha^2}(\pi(x))) = \tilde{\alpha}^2_\gamma(\tilde{\pi}(\pi_{\alpha^1}(x))).)$$

It follows that $\tilde{\pi}$ establishes an isomorphism between the dynamical systems $(M_i \otimes_{\alpha^i} G, \Gamma, \tilde{\alpha}^i)$, and the second assertion of the lemma is an immediate consequence of Prop. 4.4.4. □

Before proceeding further, let us pause to fix some notation which will minimize the cumbersomeness of the expressions to be handled: we shall work with a dynamical system $(M,G,\alpha)$, where $M \subseteq \mathfrak{L}(\mathfrak{H})$; we shall write

$$\breve{M} = M \otimes_\alpha G \subseteq \mathfrak{L}(\breve{\mathfrak{H}}),\quad \breve{\mathfrak{H}} = \mathfrak{H} \otimes L^2(G) \cong L^2(G;\mathfrak{H});$$

$$\breve{\breve{M}} = \breve{M} \otimes_{\tilde{\alpha}} \Gamma \subseteq \mathfrak{L}(\breve{\breve{\mathfrak{H}}}),\quad \breve{\breve{\mathfrak{H}}} = \breve{\mathfrak{H}} \otimes L^2(\Gamma) \cong \mathfrak{H} \otimes L^2(G) \otimes L^2(\Gamma)$$
$$\cong L^2(G \times \Gamma, \mathfrak{H}).$$

**Theorem 4.4.8.** $\breve{\breve{M}} \cong M \otimes \mathfrak{L}(L^2(G))$, *where, of course,* $M \otimes \mathfrak{L}(L^2(G))$ *is the von Neumann algebra of operators on* $\mathfrak{H} \otimes L^2(G)$ *generated by* $\{x \otimes y\colon x \in M,\ y \in \mathfrak{L}(L^2(G))\}$.

**Proof.** In view of Lemma 4.4.7 and Example 4.4.2 (b), we may assume that there is a strongly continuous unitary representation $t \to u_t$ of $G$ in $\mathfrak{H}$ such that $\alpha_t(x) = u_t x u_t^*$ for $x$ in $M$ and $t$ in $G$. We shall establish the required isomorphism via a sequence of intermediate isomorphisms.

**Step 1.** $\breve{\breve{M}} \cong M_1 \subseteq \mathfrak{L}(\breve{\breve{\mathfrak{H}}}) = \mathfrak{L}(\mathfrak{H} \otimes L^2(G) \otimes L^2(\Gamma))$, where

$$M_1 = \{x \otimes 1 \otimes 1,\ u_t \otimes \lambda_t \otimes 1,\ 1 \otimes v_\gamma \otimes \lambda_\gamma\colon x \in M,\ t \in G,\ \gamma \in \Gamma\}'',$$

where of course, $\gamma \to \lambda_\gamma$ denotes the left-regular repesentation of $\Gamma$ in $L^2(\Gamma)$, the Haar measure on $\Gamma$ so normalized that the Fourier-Plancherel transform $f \to \hat{f}$ is a unitary operator from $L^2(G)$ onto $L^2(\Gamma)$.

**Proof.** Since $\{\tilde{\alpha}_\gamma\}$ is unitarily implemented by $\{v(\gamma)\}$, it follows from Lemma 4.4.5 that if $\tilde{w}$ is the unitary operator on $\breve{\breve{\mathfrak{H}}} = L^2(\Gamma;\breve{\mathfrak{H}})$ defined by

$$(\tilde{w}\,\tilde{\tilde{\xi}}\,)(\gamma) = v(\gamma)\tilde{\tilde{\xi}}(\gamma),$$

then $\breve{\breve{M}} \cong \tilde{w}\breve{\breve{M}}\,\tilde{w}^* = M_0$ (say), where

$$M_0 = \{\tilde{x} \otimes 1_{L^2(\Gamma)}, v(\gamma) \otimes \lambda_\gamma : \tilde{x} \in \tilde{M}, \gamma \in \Gamma\}'' \subseteq \mathfrak{L}(\tilde{\mathcal{H}} \otimes L^2(\Gamma)).$$

Now if $w$ is the unitary operator on $\tilde{\mathcal{H}}$ given by $(w\tilde{\xi})(t) = u_t\tilde{\xi}(t)$, then, again by Lemma 4.4.5,

$$w\tilde{M}w^* = \{x \otimes 1_{L^2(G)}, u_t \otimes \lambda_t : x \in M, t \in G\}'' \subseteq \mathfrak{L}(\mathcal{H} \otimes L^2(G)).$$

Conclude that if

$$M_1 = (w \otimes 1_{L^2(\Gamma)})M_0(w \otimes 1_{L^2(\Gamma)})^* \subseteq \mathfrak{L}(\tilde{\mathcal{H}} \otimes L^2(\Gamma)),$$

then, since $w \in \{v(\gamma): \gamma \in \Gamma\}'$ (as is easily verified),

$$\tilde{\tilde{M}} \cong M_1 = \{x \otimes 1 \otimes 1, u_t \otimes \lambda_t \otimes 1, 1 \otimes v_\gamma \otimes \lambda_\gamma : x \in M, t \in G, \gamma \in \Gamma\}''$$
$$\subseteq \mathfrak{L}(\mathcal{H} \otimes L^2(G) \otimes L^2(\Gamma)).$$

**Step 2.** $M_1 \cong M_2 \subseteq \mathfrak{L}(\mathcal{H} \otimes L^2(G) \otimes L^2(G))$, where

$$M_2 = \{x \otimes 1 \otimes 1, u_t \otimes \lambda_t \otimes 1, 1 \otimes v_\gamma \otimes v_\gamma : x \in M, t \in G, \gamma \in \Gamma\}''.$$

**Proof.** Let $\mathcal{F}: L^2(\Gamma) \to L^2(G)$ be the Fourier-Plancherel transform (on identifying $G$ with the dual of $\Gamma$); thus $(\mathcal{F}\xi)(t) = \int \langle t,\gamma\rangle^{-1}\xi(\gamma)d\gamma$ if $\xi \in L^2(\Gamma) \cap L^1(\Gamma)$. The following facts are basic (and easily established): $\mathcal{F}\lambda_\gamma\mathcal{F}^* = v_\gamma$ and $\mathcal{F}v_t\mathcal{F}^* = \rho_t\ (= \lambda_{-t})$ for $\gamma$ in $\Gamma$ and $t$ in $G$, where of course $v_t$ denotes the multiplication operator on $L^2(\Gamma)$: $(v_t\xi)(\gamma) = \langle t,\gamma\rangle^{-1}\xi(\gamma)$. It follows from the unitarity of $\mathcal{F}$ that

$$\tilde{\tilde{\mathcal{F}}} = 1 \otimes 1 \otimes \mathcal{F}: \mathcal{H} \otimes L^2(G) \otimes L^2(\Gamma) \to \mathcal{H} \otimes L^2(G) \otimes L^2(G)$$

is unitary; it is easily deduced from the above-mentioned "basic facts" about $\mathcal{F}$ that $\tilde{\tilde{\mathcal{F}}} M_1 \tilde{\tilde{\mathcal{F}}}^* = M_2$.

**Step 3.** $M_2 \cong M_3 \subseteq \mathfrak{L}(\mathcal{H} \otimes L^2(G))$, where

$$M_3 = \{x \otimes 1, u_t \otimes \lambda_t, 1 \otimes v_\gamma : x \in M, t \in G, \gamma \in \Gamma\}''.$$

**Proof.** Identify $L^2(G) \otimes L^2(G)$ with $L^2(G \times G)$ and regard $M_2$ as a von Neumann algebra of operators on $\mathcal{H} \otimes L^2(G \times G)$. It is clear that the equation $(u\xi)(s,t) = \xi(s+t,t)$ defines a unitary operator on $L^2(G \times G)$. A minor computation reveals that, with the above identification, the following equations hold: $u^*(v_\gamma \otimes v_\gamma)u = v_\gamma \otimes 1$ and $u^*(\lambda_t \otimes 1)u = \lambda_t \otimes 1$, $\gamma \in \Gamma$, $t \in G$. (Reason:

$$((v_\gamma \otimes v_\gamma)\xi)(s,t) = \langle s+t,\gamma\rangle^{-1}\xi(s,t),$$

$$((\lambda_t \otimes 1)\xi)(s,s') = \xi(s-t,s').)$$

Conclude that

$$(1 \otimes u)^* M_2 (1 \otimes u) = \{x \otimes 1 \otimes 1,\ u_t \otimes \lambda_t \otimes 1,\ 1 \otimes v_\gamma \otimes 1:$$
$$x \in M,\ t \in G,\ \gamma \in \Gamma\}''$$
$$= M_3 \otimes 1 \cong M_3.$$

**Step 4.** $M_3 \cong M \otimes \mathfrak{L}(L^2(G))$.

**Proof.** Again, let $w$ be the unitary operator on $\mathcal{H}$ as in Lemma 4.4.5. We shall show that $w^* M_3 w = M \otimes \mathfrak{L}(L^2(G))$. Let $N = \{v_\gamma: \gamma \in \Gamma\}''$; then $w^* M_3 w = w^*((M \otimes N) \cup \{u_t \otimes \lambda_t: t \in G\})'' w$, where, as in Example 2.3.7 (b), $M \otimes N = \{x \otimes 1,\ 1 \otimes v_\gamma: x \in M,\ \gamma \in \Gamma\}''$. Suppose we can show that $w^*(M \otimes N)w = M \otimes N$; we would then have, in view of Lemma 4.4.5, $w^* M_3 w = ((M \otimes N) \cup \lambda(G))'' = M \otimes (N \cup \{\lambda_t: t \in G\})''$; it is however a standard fact that only scalar operators commute with all multiplications (i.e., members of $N$) and all translations (i.e., members of $\{\lambda_t: t \in G\}''$); in other words, $(N \cup \{\lambda_t: t \in G\})'' = \mathfrak{L}(L^2(G))$ and the proof will be complete.

To prove $w^*(M \otimes N)w = M \otimes N$, we shall rely fairly heavily on the fact (mentioned in Example 2.3.7 (b)) that $(M \otimes N)' = M' \otimes N'$. Before proceeding with the proof, let us first obtain an easy consequence of the above fact: if $M_i \subseteq \mathfrak{L}(\mathcal{H})$ and $N_i \subseteq \mathfrak{L}(\quad)$ are von Neumann algebras, for $i = 1,2$, then

$$((M_1 \otimes N_1) \cup (M_2 \otimes N_2))'' = (M_1 \cup M_2)'' \otimes (N_1 \cup N_2)''$$

by definition of the tensor product of von Neumann algebras; it follows from the above fact, the preceding fact concerning commutants of tensor products, and the double commutant theorem that $(M_1 \otimes N_1) \cap (M_2 \otimes N_2) = (M_1 \cap M_2) \otimes (N_1 \cap N_2)$. (Verify this, using Ex. (0.4.15) (b)!) After this lengthy preamble, let us get on with the proof.

It is a standard fact from harmonic analysis that $N$ is maximal abelian: i.e., $N = N'$. Consequently, by Lemma 4.4.5 and Lemma 4.4.6 (b),

$$w^*(M \otimes 1)w = \pi_\alpha(M) \subseteq (1 \otimes N)' = \mathfrak{L}(\mathcal{H}) \otimes N; \tag{4.4.2}$$

on the other hand, (since every operator in $M' \otimes 1$ is of the form $(\tilde{x}'\xi)(s) = x'\xi(s)$ for some $x'$ in $M'$) it is easily verified that

$$\pi_\alpha(M) \subseteq (M' \otimes 1)' = M \otimes \mathfrak{L}(L^2(G)); \tag{4.4.3}$$

it is also easily verified that $w^* v(\gamma) w = v(\gamma)$ for all $\gamma$ in $\Gamma$, and hence,

$$w^*(1 \otimes N)w = 1 \otimes N \subseteq M \otimes N; \tag{4.4.4}$$

it follows from our preliminary remark above (concerning intersections of tensor products of von Neumann algebras) and equations (4.4.2), (4.4.3) and (4.4.4) that

$$w^*(M \otimes N)w \subseteq M \otimes N.$$

For the reverse inclusion, begin by observing -- after an easy computation -- that if $x \in M$ and $\xi \in \mathcal{H}$, then $(w(x \otimes 1)w^*\xi)(s) = \alpha_s(x)\xi(s)$. If $J$ is the unitary operator on $L^2(G)$ given by $(J\xi)(s) = \xi(-s)$, the previous equation says that $w(x \otimes 1)w^* = (1 \otimes J)\pi(x)(1 \otimes J) \in M \otimes \mathfrak{L}(L^2(G))$, and so, $w(M \otimes 1)w^* \subseteq M \otimes \mathfrak{L}(L^2(G))$. Also, notice -- from the above formula for $w(x \otimes 1)w^*$ -- that $w(M \otimes 1)w^* \subseteq (1 \otimes N)' = \mathfrak{L}(\mathcal{H}) \otimes N$, so that, as before, $w(M \otimes 1)w^* \subseteq M \otimes N$ and $w(1 \otimes N)w^* = 1 \otimes N \subseteq M \otimes N$, whence $w(M \otimes N)w^* \subseteq M \otimes N$, and the proof is finally complete. □

## Exercises

**(4.4.9)** Let $(M,G,\alpha)$, $\{u_t\}$ be as in Theorem 4.4.8.

(a) Show that there is an action $\theta$ of $G$ on $M \otimes \mathfrak{L}(L^2(G))$ such that $\theta_t = \mathrm{ad}(u_t \otimes \rho_t)$ for all $t$ -- i.e., $\theta_t(x) = (u_t \otimes \rho_t)x(u_t \otimes \rho_t)^*$ for $x$ in $M \otimes \mathfrak{L}(L^2(G))$ and $t$ in $G$ -- where, as usual, $t \to \rho_t$ denotes the right regular representation of $G$ in $L^2(G)$.

(b) Since $G$ may be identified with the dual group of $\Gamma$, the action $\tilde{\alpha}$ of $\Gamma$ on $M$ induces a dual action $\tilde{\tilde{\alpha}}$ of $G$ on $\tilde{M}$, as in Lemma 4.4.6. Show that the dynamical systems $(\tilde{M},G,\tilde{\tilde{\alpha}})$ and $(M \otimes \mathfrak{L}(L^2(G)),G,\theta)$ are isomorphic, where $\theta$ is as in (a). (Hint: under the isomorphisms $\tilde{M} \cong M_1 \cong M_2 \cong M_3 \cong M \otimes \mathfrak{L}(L^2(G))$ established in the proof of Theorem 4.4.8, let the action $\tilde{\tilde{\alpha}}$ be successively transformed into $\alpha^1$, $\alpha^2$, $\alpha^3$ and $\alpha^4$; writing ad $v$ for the automorphism $y \to vyv^*$, show that $\alpha^1_t = \mathrm{ad}(1 \otimes 1 \otimes v_t)$, $\alpha^2_t = \mathrm{ad}(1 \otimes 1 \otimes \rho_t)$, $\alpha^3_t = \mathrm{ad}(1 \otimes \rho_t)$ and $\alpha^4_t = \mathrm{ad}(u_t \otimes \rho_t) = \theta_t$.) □

**Definition 4.4.10.** A von Neumann algebra $M$ is said to be properly infinite if there is a sequence $\{e_n\}_{n=1}^{\infty}$ of pairwise orthogonal projections in $M$ such that $\Sigma e_n = 1$ and $e_n \sim 1$ (rel $M$) for all $n$. □

Observe that (by the proof of Corollary 1.2.4 (a)) every infinite factor (i.e., of type $\mathrm{I}_\infty$, $\mathrm{II}_\infty$ or III) is properly infinite.

**Lemma 4.4.11.** *If $M$ is properly infinite, and if $\mathcal{H}$ is any separable Hilbert space, them $M \otimes \mathfrak{L}(\mathcal{H}) \cong M$.*

**Proof.** By hypothesis, there exists a sequence of pairwise orthogonal projections $e_n$ in $M$ and partial isometries $v_n$ in $M$ such that $\Sigma e_n = 1$, $v_n^* v_n = e_n$ and $v_n v_n^* = 1$ for all $n$. Notice that $v_m v_n^* = \delta_{mn} 1$ for all $m$ and $n$.

Let $\{e_{ij}\}$ be matrix units in $\mathfrak{L}(\mathcal{H})$ -- i.e., pick an orthonormal basis

$\{\xi_j\}$ for $\mathcal{H}$ and let $e_{ij} = t_{\xi_i,\xi_j}$. For $k = 1,2, \ldots$, define

$$u_k = \sum_{i=1}^{k} v_i \otimes e_{i1} \in M \otimes \mathfrak{L}(\mathcal{H}),$$

and observe that

$$u_k^* u_k = \sum_{i,j=1}^{k} (v_i^* v_j) \otimes (e_{1i} e_{j1}) = \sum_{i=1}^{k} v_i^* v_i \otimes e_{11}$$

$$= \left[\sum_{i=1}^{k} e_i\right] \otimes e_{11},$$

with

$$u_k u_k^* = \sum_{i,j=1}^{k} (v_i v_j^*) \otimes (e_{i1} e_{1j}) = \sum_{i=1}^{k} 1 \otimes e_{ii} = 1 \otimes \left[\sum_{i=1}^{k} e_{ii}\right].$$

Identical computations show that if $k > \ell$,

$$(u_k^* - u_\ell^*)(u_k - u_\ell) = \left[\sum_{\ell < i \leq k} e_i\right] \otimes e_{11},$$

while

$$(u_k - u_\ell)(u_k^* - u_\ell^*) = 1 \otimes \left[\sum_{\ell < i \leq k} e_{ii}\right].$$

Since

$$\sum_{i=1}^{k} e_i \to 1 \quad \text{and} \quad \sum_{i=1}^{k} e_{ii} \to 1$$

strongly (in fact $\sigma$-strongly), conclude that $\{u_k\}$ converges strongly* (in fact, $\sigma$-strongly*) to a $u \in M \otimes \mathfrak{L}(\mathcal{H})$, where $u^*u = 1 \otimes e_{11}$ and $uu^* = 1 \otimes 1$. So $u^*$ is isometric; it follows that the map $x \to u^*xu$ defines an isomorphism of $M \otimes \mathfrak{L}(\mathcal{H})$ onto $M \otimes e_{11}$, which, in turn, is isomorphic to $M$. □

Combining Theorem 4.4.8 and Lemma 4.4.11, we get the following duality theorem.

**Proposition 4.4.12.** *Let $\alpha$ be an action of a locally compact abelian group $G$ on a properly infinite von Neumann algebra $M$. If $\tilde{\alpha}$ denotes the dual action of $\Gamma$ on $\tilde{M}$ (as in Lemma 4.4.6), then $M \cong \tilde{\tilde{M}}$, where, as before, $\tilde{\tilde{M}} = (M \otimes_\alpha G) \otimes_{\tilde{\alpha}} \Gamma$.*

## 4.5. The Structure of Properly Infinite von Neumann Algebras

Recall that actions $\alpha$ and $\beta$ of a locally compact group $G$ on $M$ are said to be outer equivalent if there is a strongly continuous map $t \to u_t$ from $G$ to $\mathcal{U}(M)$ such that $u_{st} = u_s\alpha_s(u_t)$ and $\beta_t(x) = u_t\alpha_t(x)u_t^*$ for all $s,t$ in $G$ and $x$ in $M$.

**Lemma 4.5.1.** *If $\alpha$ and $\beta$ are outer equivalent actions of a locally compact group $G$ on a von Neumann algebra $M$, then $M \otimes_\alpha G \cong M \otimes_\beta G$.*

**Proof.** Assume $M \subseteq \mathfrak{L}(\mathcal{H})$, so that both the relevant von Neumann algebras act on $\tilde{\mathcal{H}} = L^2(G;\mathcal{H})$. It is easily checked that the equation $(\tilde{u}\xi)(t) = u_{t^{-1}}\xi(t)$ defines a unitary operator $\tilde{u}$ on $\tilde{\mathcal{H}}$, with inverse given by $(\tilde{u}^*\xi)(t) = u^*_{t^{-1}}\xi(t)$. Let $x \in M$, $t \in G$; then

$$\begin{aligned}(\tilde{u}\pi_\alpha(x)\tilde{u}^*\xi)(s) &= u_{s^{-1}}(\pi_\alpha(x)\tilde{u}^*\xi)(s) = u_{s^{-1}}\alpha_{s^{-1}}(x)(\tilde{u}^*\xi)(s)\\ &= u_{s^{-1}}\alpha_{s^{-1}}(x)u^*_{s^{-1}}\xi(s) = \beta_{s^{-1}}(x)\xi(s)\\ &= (\pi_\beta(x)\xi)(s);\end{aligned}$$

$$\begin{aligned}(\tilde{u}\lambda(t)\tilde{u}^*\xi)(s) &= u_{s^{-1}}(\tilde{u}^*\xi)(t^{-1}s)\\ &= u_{s^{-1}}u^*_{s^{-1}t}\xi(t^{-1}s)\\ &= u_{s^{-1}}(u_{s^{-1}}\alpha_{s^{-1}}(u_t))^*\xi(t^{-1}s)\\ &= u_{s^{-1}}\alpha_{s^{-1}}(u_t^*)u^*_{s^{-1}}\xi(t^{-1}s)\\ &= \beta_{s^{-1}}(u_t^*)(\lambda(t)\xi)(s)\\ &= (\pi_\beta(u_t^*)\lambda(t)\xi)(s);\end{aligned}$$

thus

$$\tilde{u}\,\pi_\alpha(x)\tilde{u}^* = \pi_\beta(x)$$

and

$$\tilde{u}\,\lambda(t)\tilde{u}^* = \pi_\beta(u_t^*)\lambda(t),$$

and hence, $\tilde{u}(M \otimes_\alpha G)\tilde{u}^* \subseteq M \otimes_\beta G$.

An entirely similar computation shows that $\tilde{u}^*(M \otimes_\beta G)\tilde{u} \subseteq M \otimes_\alpha G$, whence $M \otimes_\alpha G$ and $M \otimes_\beta G$ are spatially isomorphic. □

**Corollary 4.5.2.** *If $\phi$ and $\psi$ are* fns *weights on a von Neumann algebra $M$, then $M \otimes_{\sigma^\phi} \mathbb{R} \cong M \otimes_{\sigma^\psi} \mathbb{R}$.*

**Proof.** This is an immediate consequence of the unitary cocycle theorem and the preceding Lemma. □

**Proposition 4.5.3.** *Let $\phi$ be a* fns *weight on $M$. Identify $M$ with $\pi_\phi(M)$ and assume that $M \subseteq \mathfrak{L}(\mathcal{H})$, $\mathcal{H} = \mathcal{H}_\phi$ and $\pi_\phi(x) = x$ for all $x$ in $M$. Let $\alpha$ be an action of a locally compact abelian group $G$ on $M$ such that $\phi \circ \alpha_t = \phi$ for all $t$. Then there exists a (canonically constructed)* fns *weight $\breve{\phi}$ on $\breve{M}$ such that $\breve{\mathcal{H}} = L^2(G;\mathcal{H})$ may be identified with $\mathcal{H}_{\breve{\phi}}$ in such a way that $\pi_{\breve{\phi}} = \mathrm{id}_{\breve{M}}$ and the following statements are valid:*

(a) $\mathrm{dom}\,\Delta_{\breve{\phi}}^{1/2} = \{\breve{\xi} \in \breve{\mathcal{H}} : \breve{\xi}(t) \in \mathrm{dom}\,\Delta_{\phi}^{1/2}$ *for all t and* $\int \|\Delta_{\phi}^{1/2}\breve{\xi}(t)\|^2 dt < \infty\}$

*and* $(\Delta_{\breve{\phi}}^{1/2}\breve{\xi})(t) = \Delta_{\phi}^{1/2}\breve{\xi}(t)$ *for* $\breve{\xi}$ *in* $\mathrm{dom}\,\Delta_{\breve{\phi}}^{1/2}$;

(b) *the dual weight $\breve{\phi}$ is invariant under the dual action $\tilde{\alpha}$ of $\Gamma$ on $\breve{M}$ -- i.e., $\breve{\phi} \circ \tilde{\alpha}_\gamma = \breve{\phi}$ for all $\gamma$ in $\Gamma$, where $\tilde{\alpha}$ is defined as in Lemma* 4.4.6.

**Proof.** As the proof of this result is quite involved and technical (cf. [Tak 3]), we shall only present the proof for discrete $G$. In the discrete case, we are back in the situation encountered in the first half of Section 4.2. Let $\breve{\phi}$ denote the fns weight on $\breve{M}$ defined (as in Section 4.2) by $\breve{\phi}(\breve{x}) = \phi(\breve{x}(\epsilon,\epsilon))$. As before, we shall write , $\eta$, $S$, $J$, $\Delta$ and $\breve{N}$, $\breve{\eta}$, $\breve{S}$, $\breve{J}$, $\breve{\Delta}$ for $N_\phi$, $\eta_\phi$, $S_\phi$, $J_\phi$, $\Delta_\phi$ and $N_{\breve{\phi}}$, $\eta_{\breve{\phi}}$, $S_{\breve{\phi}}$, $J_{\breve{\phi}}$, $\Delta_{\breve{\phi}}$

respectively. Let us also write $\mathcal{U}$ and $\breve{\mathcal{U}}$ for $N \cap N^*$ and $\breve{N} \cap \breve{N}^*$ respectively.

It follows from the assumption $\phi \circ \alpha_t = \phi$ that $\alpha_t(N) \subseteq N$ and that there is a unitary representation $t \to u_t$ of $G$ in $\mathcal{H}$ such that $u_t\eta(x) = \eta(\alpha_t(x))$ for $x$ in $N$ and $t$ in $G$. If $\breve{x} \in \breve{M}$, recall (cf. Prop. 4.2.2 (a)) that $\breve{x} \in \breve{N} \Leftrightarrow \breve{x}(t,\epsilon) \in N$ $\forall t$ in $G$ and

$$\sum_{t\in G} \|\eta(\breve{x}(t,\epsilon))\|^2 < \infty ;$$

also

$$\breve{x}^*(t,\epsilon) = \breve{x}(\epsilon,t)^* = \alpha_{t^{-1}}(\breve{x}(t^{-1},\epsilon)^*);$$

since $\|\eta(\alpha_t(x))\| = \|\eta(x)\|$ for $x$ in $N$ and $t$ in $G$, deduce that

$$\breve{\mathcal{U}} = \{\breve{x} \in \breve{H}: \breve{x}(t,\epsilon) \in \mathcal{U} \quad \forall t \text{ in } G, \text{ and}$$
$$\sum_{t\in G} (\|\eta(\breve{x}(t,\epsilon))\|^2 + \|\eta(\breve{x}(t,\epsilon)^*)\|^2) < \infty\}.$$

If $\breve{\mathfrak{D}}_0 = \breve{\eta}(\breve{\mathcal{U}})$, we thus have

$$\breve{\mathfrak{D}}_0 = \{\breve{\xi} \in \breve{\mathcal{H}}: \breve{\xi}(t) \in \eta(\mathcal{U}) \quad \forall t \text{ in } G, \text{ and}$$
$$\sum_{t\in G} (\|\breve{\xi}(t)\|^2 + \|S\breve{\xi}(t)\|^2) < \infty\}.$$

By the definition of $\breve{S}$, $\breve{\mathfrak{D}}_0$ is a core for $\breve{S}$; on the other hand, since $\|S\xi\|^2 = \|\Delta^{1/2}\xi\|^2$ for $\xi$ in $\mathcal{U}$, and since (again by definition) $\eta(\mathcal{U})$ is a core for $S$ (and consequently for $\Delta^{1/2}$), it follows from Ex. (2.5.6)(b) that $\breve{\mathfrak{D}}_0$ is also a core for the positive self-adjoint operator $H$ in $\breve{\mathcal{H}}$ given by $H = \oplus H_t$ (in the sense of Ex. (2.5.6)), where $H_t = \Delta^{1/2}$ for all $t$.

It is clear that the equation $(\hat{J}\breve{\xi})(t) = u_{t^{-1}}J\breve{\xi}(t^{-1})$ defines an antiunitary operator on $\breve{\mathcal{H}}$; observe that if $\tilde{x} \in \breve{U}$, then

$$\begin{aligned}(\hat{J}\,\breve{H}\,\breve{\eta}(\tilde{x}))(t) &= u_{t^{-1}}J(\breve{H}\,\breve{\eta}(\tilde{x}))(t^{-1}) \\ &= u_{t^{-1}}J\Delta^{1/2}(\breve{\eta}(\tilde{x})(t^{-1})) \\ &= u_{t^{-1}}\,J\Delta^{1/2}\eta(\tilde{x}(t^{-1},\epsilon)) \\ &= u_{t^{-1}}\eta(\tilde{x}(t^{-1},\epsilon)^*) \\ &= \eta(\alpha_{t^{-1}}(\tilde{x}(t^{-1},\epsilon)^*)) \\ &= \eta(\tilde{x}(\epsilon,t)^*) \\ &= \eta(\tilde{x}^*(t,\epsilon) \\ &= \breve{\eta}(\tilde{x}^*)(t).\end{aligned}$$

Thus $\hat{J}\breve{H}\,\breve{\xi} = \breve{S}\,\breve{\xi}$ whenever $\breve{\xi} \in \breve{\mathfrak{D}}_0$. Since $\breve{\mathfrak{D}}_0$ is a core for $\breve{H}$ as well as for $\breve{S}$, and since $\hat{J}$ is antiunitary, deduce that $\breve{S} = \hat{J}\,\breve{H}$. Infer from the uniqueness of the polar decomposition that $\hat{J} = \breve{J}$ and $\breve{H} = \breve{\Delta}^{1/2}$.

For (b), notice that if $\gamma \in \Gamma$, then

$$v(\gamma) = \bigoplus_{t\in G} \langle t,\gamma\rangle^{-1}1_{\mathcal{H}}\,;$$

if $\tilde{x} \in \breve{M}$, an easy matrix multiplication shows that for all $s,t$ in $G$, $(\tilde{\alpha}_\gamma(\tilde{x}))(s,t) = \langle s,\gamma\rangle^{-1}\langle t,\gamma\rangle\tilde{x}(s,t)$, and in particular, $(\tilde{\alpha}_\gamma(\tilde{x}))(\epsilon,\epsilon) = \tilde{x}(\epsilon,\epsilon)$; it follows from the definition of $\phi$ that $\phi \circ \tilde{\alpha}_\gamma = \phi$. (Note: The commutativity of $G$ is not used anywhere in this proof of the discrete case of (a).) □

## Exercises

**(4.5.4)** Let $(M,G,\alpha)$ and $\phi$ be as in Theorem 4.5.3. Assume $G$ is discrete. If $u_t$ is as in the proof of Theorem 4.5.3, and if $u_t \in M$, show that $u_t \in Z(M)$. (Hint: use the formula obtained for $\breve{J}$ (= $\hat{J}$) and the fact that $\breve{J}^2 = 1$ to conclude that $Ju_tJ = u_t$; now appeal to the Tomita-Takesaki theorem.) □

**Theorem 4.5.5.** *If $M$ is properly infinite (cf. Definition 4.4.10), there exists a properly infinite but semifinite von Neumann algebra $N$ and an action $\theta$ of $\mathbb{R}$ on $N$ such that $M \cong N \otimes_\theta \mathbb{R}$. Further, there exists a* fns *trace $\tau$ on $N$ such that $\tau \circ \theta_t = e^{-t}\tau$ for all $t$ in $\mathbb{R}$.*

**Proof.** Let $\phi$ be a fns weight on $M$, assume $M \subseteq \mathcal{L}(\mathcal{H})$, $\mathcal{H} = \mathcal{H}_\phi$ and $\pi_\phi =$

$\mathrm{id}_M$. Let $N = \check{M} = M \otimes_{\sigma^\phi} \mathbb{R}$, and let $\theta$ denote the dual action $(\sigma^\phi)^\sim$ of (the dual group of $\mathbb{R}$ which we identify with) $\mathbb{R}$ on $N$. Then, by Proposition 4.4.12, we have $M \cong N \otimes_\theta \mathbb{R}$.

Since $M$ is properly infinite, so is $\pi(M) \subseteq N$, and consequently $N$ is also properly infinite.

Let $\check{\phi}$ denote the dual weight on $N$ (as in Proposition 4.5.3). We shall show that for any $\tilde{x}$ in $N$ and $t$ in $\mathbb{R}$,

$$(*) \qquad \sigma_t^{\check{\phi}}(\tilde{x}) = \lambda(t)\, \tilde{x}\, \lambda(t)^*;$$

since $\lambda(t) \in N$, this would say that the flow $\sigma^{\tilde{\phi}}$ is inner, and the semifiniteness of $N$ would follow from Theorem 3.1.6.

In order to establish (*), we may assume, without loss of generality, that $\tilde{x} \in \pi(M) \cup \lambda(\mathbb{R})$. If $x \in M$ and $s,t,u \in \mathbb{R}$, compute as follows:

$$\begin{aligned}
(\sigma_t^{\check{\phi}}(\pi(x))\check{\xi})(s) &= (\Delta_{\check{\phi}}^{it}\, \pi(x) \Delta_{\check{\phi}}^{-it}\, \check{\xi})(s) \\
&= \Delta_{\phi}^{it}(\pi(x)\Delta_{\check{\phi}}^{-it}\, \check{\xi})(s) \qquad \text{(by Prop. 4.5.3)} \\
&= \Delta_{\phi}^{it}\, \sigma_{-s}^{\phi}(x)(\Delta_{\check{\phi}}^{-it}\, \check{\xi})(s) \\
&= \Delta_{\phi}^{it}\, \sigma_{-s}^{\phi}(x)\Delta_{\phi}^{-it}\, \check{\xi}(s) \qquad \text{(again, by Prop. 4.5.3)} \\
&= \sigma_{t-s}^{\phi}(x)\check{\xi}(s) \\
&= (\pi(\sigma_t^{\phi}(x))\check{\xi})(s),
\end{aligned}$$

so that

$$\begin{aligned}
\sigma_t^{\check{\phi}}(\pi(x)) &= \pi(\sigma_t^{\phi}(x)) \\
&= \lambda(t)\pi(x)\lambda(t)^* \qquad \text{(by equation (4.4.1));}
\end{aligned}$$

$$\begin{aligned}
(\sigma_t^{\check{\phi}}(\lambda(u))\check{\xi})(s) &= (\Delta_{\check{\phi}}^{it}\, \lambda(u)\Delta_{\check{\phi}}^{-it}\, \check{\xi})(s) \\
&= \Delta_{\phi}^{it}(\lambda(u)\Delta_{\check{\phi}}^{-it}\, \check{\xi})(s) \\
&= \Delta_{\phi}^{it}(\Delta_{\check{\phi}}^{-it}\, \check{\xi})(s - \mathrm{u}) \\
&= \Delta_{\phi}^{it}\Delta_{\phi}^{-it}\, \check{\xi}(s - u) = \check{\xi}(s - u),
\end{aligned}$$

so that $\sigma_t^{\check{\phi}}(\lambda(u)) = \lambda(u) = \lambda(t)\lambda(u)\lambda(t)^*$; this completes the proof of semifiniteness of $N$.

By Stone's theorem, there exists an invertible positive self-adjoint operator $H$ affiliated to $N$ such that $\lambda(t) = \check{H}^{it}$ for all $t$ in $\mathbb{R}$. Since

$$\sigma_t^{\check{\phi}}(\tilde{x}) = \check{H}^{it}\, \tilde{x}\, \check{H}^{-it}$$

for $\tilde{x}$ in $N$ and $t$ in $\mathbb{R}$, it follows at once that $\widetilde{H}\ \eta\ N^{\widetilde{\phi}}$. Hence, the equation $\tau = \phi(H^{-1}\cdot)$ defines a fns weight $\tau$ on $N$. According to Theorem 3.1.10, we have

$$\sigma_t^\tau(\tilde{x}) = \widetilde{H}^{-it}\sigma_t^{\widetilde{\phi}}(\tilde{x})\widetilde{H}^{it} = \tilde{x}$$

for all $\tilde{x}$ in $N$, and consequently $\tau$ is a trace.

Observe next that if $s,t \in \mathbb{R}$, then $\theta_s(\lambda(t)) = v(s)\lambda(t)v(-s) = e^{-ist}\lambda(t)$ [cf. Lemma 4.4.6 (a)]; it follows -- by "passing to the infinitesimal generator" -- that $\theta_s(H^{-1}) = e^s H^{-1}$. (The last equation is meaningful if $H^{-1}$ is bounded; otherwise it must be interpreted as $\theta_s(1_E(H^{-1})) = 1_E(e^s H^{-1})$ for every Borel set $E$ in $\mathbb{R}$.) Hence, if $\tilde{x} \in N_+$,

$$\begin{aligned}\tau \circ \theta_s(\tilde{x}) &= \widetilde{\phi}(\widetilde{H}^{-1}\theta_s(\tilde{x}))\\ &= \widetilde{\phi} \circ \theta_s(e^{-s}\,\widetilde{H}^{-1}\,\tilde{x})\\ &= e^{-s}\,\widetilde{\phi}(\widetilde{H}^{-1}\,\tilde{x}) \qquad \text{(by Prop. 4.5.3 (b))}\\ &= e^{-s}\,\tau(\tilde{x}),\end{aligned}$$

where, for convenience, we have written expressions such as $\phi(H^{-1}\theta_s(\tilde{x}))$ in place of the more accurate (but more cumbersome)

$$\lim_{\epsilon \downarrow 0}\ \widetilde{\phi}((\widetilde{H}^{-1})_\epsilon\ \theta_s(\tilde{x}));$$

the proof is complete. □

It is a fact that if $\alpha$ is an action of a locally compact abelian group $G$ on $M$, then $\pi_\alpha(M)$ is precisely the fixed point algebra $\widetilde{M}^{\widetilde{\alpha}}$. It can be deduced from this that the dynamical system $(N, \mathbb{R}, \theta)$ of Theorem 4.5.5 is unique in the following sense: if $(N_i, \mathbb{R}, \theta^i)$ $i = 1,2$ are dynamical systems, if $\tau_i$ is a fns trace on $N_i$ such that $\tau_i \circ \theta_t^i = e^{-t}\tau_i$ for all $t$ in $\mathbb{R}$ and $i = 1,2$, and if the crossed-products $N_1 \otimes_{\theta^1} \mathbb{R}$ and $N_2 \otimes_{\theta^2} \mathbb{R}$ are isomorphic von Neumann algebras, then there exists a von Neumann algebra isomorphism $\pi\colon N_1 \to N_2$ such that if $\theta_t(x) = \pi^{-1}(\theta_t^2(\pi(x))$ for $x$ in $N_1$ and $t$ in $\mathbb{R}$, then the actions $\theta$ and $\theta^1$ of $G$ on $N_1$ are outer equivalent. The details of this argument, and, in fact, the entire discussion of Sections 4.4 and 4.5 may be found in [Tak 3].

Since a major portion of this book has been devoted to factors, it is only fitting that we conclude with the statement of two very beautiful and powerful structure theorems:

(1) Let $M$, $N$, $\theta$ and $\tau$ be as in Theorem 4.5.5. Then $M$ is a factor of type $\mathrm{III}_1$ if and only if $N$ is a factor of type $\mathrm{II}_\infty$;

(2) If $M$ is a factor of type $\mathrm{III}_\lambda$, $0 < \lambda < 1$, there exists a factor $N$ of type $\mathrm{II}_\infty$ and an automorphism $\alpha_1$ of $N$ such that $\tau \circ \alpha_1 = \lambda\tau$, where $\tau$ is the essentially unique fns trace on $N$, such that $M \cong N \otimes_\alpha \mathbb{Z}$, where $\alpha_n = \alpha_1^n$. □

# APPENDIX
# TOPOLOGICAL GROUPS

A topological group is a pair $(G,\tau)$ where $\tau$ is a Hausdorff topology on a group $G$ such that the maps $G \times G \to G$ and $G \to G$ defined by $(s,t) \to st$ and $t \to t^{-1}$ are continuous. As is customary, we shall not explicitly refer to $\tau$, and simply call $G$ a topological group. Since left- (resp. right-) translations of $G$ -- i.e. the maps $t \to st$ (resp. $t \to ts$) for a fixed $s$ in $G$ -- are homeomorphisms of $G$ onto itself, a topological group is homogeneous in the sense that its group of homeomorphisms acts transitively. An easy consequence of this observation is that a homomorphism $f$: $G_1 \to G_2$ of topological groups is continuous if and only if it is continuous at the identity element; similarly, a topological group is locally compact if and only if there exists a compact neighborhood of the identity element.

Assume henceforth that $G$ is a locally compact group, which, for convenience, will be assumed to have a countable base of open sets. Let $\mathcal{F}_G$ denote the $\sigma$-algebra generated by compact subsets of $G$. The assumed second countability ensures that every open set is a countable union of compact sets, and hence an element of $\mathcal{F}_G$. (For groups that are too "large", one must distinguish between Baire sets and Borel sets; we shall ignore such pathological groups; the interested reader may find such a treatment in [Hal 1] or [Loo].)

The basic fact concerning locally compact groups is this: there exists a positive measure $\mu$ defined on $\mathcal{F}_G$ such that (i) $\mu(K) < \infty$ for every compact subset $K$ of $G$; (ii) $\mu(U) > 0$ for every non-empty open set $U$; and (iii) $\mu(E) = \mu(sE)$ $(= \mu(\{st: t \in E\}))$ for every $E$ in $\mathcal{F}_G$. If $\nu$ is another measure on $\mathcal{F}_G$ satisfying the three conditions above, then there is a positive real number $c$ such that $\nu = c\mu$. A measure as above is called a left Haar measure on $G$, the qualifier "left" stemming from the fact that we have only required invariance under left translations.

If $\mu_\ell$ is a left Haar measure for $G$, it is easily verified that the equation $\mu_r(E) = \mu_\ell(E^{-1})$ (where $E^{-1} = \{t^{-1}: t \in E\}$) defines a right Haar measure $\mu_r$ on $G$; it follows that $G$ admits a right Haar measure which is unique up to constant multiples.

It may, however, be the case that the left Haar measure is not a right Haar measure. In such a case, something can still be salvaged. Let $\mu$ be a left Haar measure. If $s \in G$, define a measure $\mu_s$ by $\mu_s(E) = \mu(Es)$; it is immediate that $\mu_s$ is also a left Haar measure and consequently there is a constant $\Delta(s) > 0$ such that $\mu(Es) = \Delta(s)\mu(E)$ for all $E$ in $\mathcal{F}_G$. The definition of $\Delta$ implies that $\Delta$ is a continuous homomorphism from $G$ into the multiplicative grup $\mathbb{R}_+^*$ of positive real numbers; the function $\Delta$ is called the modular function of $G$; it is characterized, in the language of integrals, by the requirement that

$$\int f(st)ds = \Delta(t^{-1}) \int f(s)ds$$

for every $f$ in $C_c(G)$.

There are two special cases when $\Delta$ is trivial: (i) $G$ is compact (since the only compact subgroup of $\mathbb{R}_+^*$ is $\{1\}$); and (ii) $G$ is abelian. A group is called unimodular if $\Delta(t) = 1$ for all $t$. Another example of a unimodular group is the group $GL(n,\mathbb{C})$.

Assume, henceforth, that $G$ is a locally compact abelian group; as is customary, we shall employ the additive notation in $G$. A character of $G$ is a continuous homomorphism from $G$ into the compact multiplicative group $T$ of complex numbers of unit modulus. It is clear that the set of characters of $G$ is an abelian group $\Gamma$ with respect to pointwise operations: if $\gamma_1,\gamma_2 \in \Gamma$ and $t \in G$, then $\langle t,\gamma_1+\gamma_2\rangle = \langle t,\gamma_1\rangle\langle t,\gamma_2\rangle$, where we write $\langle t,\gamma\rangle$ instead of $\gamma(t)$.

The group $\Gamma$, being a function space, may be equipped with the so-called "compact-open" topology: a typical sub-basic open neighborhood of a point $\gamma_0$ in $\Gamma$ is given by $W(K,U) = \{\gamma \in \Gamma: \gamma(K) \subseteq U\}$, where $K$ is a compact set in $G$ and $U$ is an open subset of $T$ such that $\gamma_0(K) \subseteq U$. It may then be shown that this topology equips $\Gamma$ with the structure of a locally compact abelian group.

Each $t$ in $G$ defines a character $\phi_t$ of $\Gamma$ by the equation $\phi_t(\gamma) = \langle t,\gamma\rangle$. Since $\Gamma$ is a locally compact abelian group, one can construct the character group $\mathcal{G}$ of $\Gamma$, which is again a locally compact abelian group. The celebrated Pontrjagin duality theorem asserts that the map $t \to \phi_t$ defines a homeomorphic isomorphism of $G$ onto $\mathcal{G}$. For this reason, $\Gamma$ is sometimes also called the dual of the group $G$.

If $f \in L^1(G)$, define the Fourier-transform of $f$ to be the function $\hat{f}: \Gamma \to \mathbb{C}$ defined by

$$\hat{f}(\gamma) = \int \langle t,\gamma\rangle^{-1} f(t)dt.$$

It may be shown that, with respect to the topology on $\Gamma$, the function $\hat{f}$ is continuous. (The Riemann-Lebesgue lemma goes one step further and states that $\hat{f}$ is a continuous function which vanishes at infinity -- i.e., is uniformly approximable by continuous functions of compact support.)

The map $f \to \hat{f}$ is, thus, a norm-decreasing homomorphism of commutative Banach algebras ($L^1(G) \to C_0(\Gamma)$). It can be shown that every non-zero multiplicative homomorphism from $L^1(G)$ to $\mathbb{C}$ is of the form $f \to \hat{f}(\gamma)$ for some $\gamma$ in $\Gamma$; thus $\Gamma$ may be identified with the maximal ideal space of the commutative Banach algebra $L^1(G)$; this identification is also a topological one: $\gamma_i \to \gamma$ in the topology of $\Gamma$ if and only if $\hat{f}(\gamma_i) \to \hat{f}(\gamma)$ for each $f$ in $L^1(G)$. Thus, the Fourier transform may also be viewed as the Gelfand transform.

A basic fact concerning Fourier transforms is the Inversion Theorem which states that it is possible to normalize the Haar measure on $\Gamma$ so that if $f \in L^1(G)$ and $\hat{f} \in L^1(G)$, then

$$f(t) = \int_\Gamma \langle t,\gamma \rangle \hat{f}(\gamma)d\gamma.$$

For instance if $G = \mathbb{R}^n$, the dual $\Gamma$ may be identified with $\mathbb{R}^n$, the duality being given by

$$\langle t,a \rangle = \exp\left[ i \sum_{j=1}^{n} t_j a_j \right].$$

If the Haar measure on $\mathbb{R}^n$, which is Lebesgue measure, is normalized so that $[0,1]^n$ receives unit measure, the Haar measure on $\Gamma = \mathbb{R}^n$ must be chosen as $(2\pi)^n$ times this measure.

An immediate consequence of the Inversion Theorem is the injectivity of the map $f \to \hat{f}$ from $L^1(G)$ to $C_0(\Gamma)$. Another fact which relies on the Inversion Theorem is Plancherel's theorem which states that there exists a unique unitary operator $\mathcal{F}$: $L^2(G) \to L^2(\Gamma)$ so that $\mathcal{F}f = \hat{f}$ for $f$ in $L^2(G) \cap L^1(G)$ -- assuming that the Haar measure on $\Gamma$ is normalized as in the Inversion Theorem.

# NOTES

(An attempt is made, in this section, to attribute credit (of a bibliographic nature) where due. If there are shortcomings, the author apologizes, in advance, for such inadvertent errors.)

## Chapter 0

The material in Section 0.1 may be found in any standard text on Hilbert space theory, such as [Hal 2] or [RS].

The treatment in Sections 0.2 and 0.3 is essentially as in [Tak 4] (Sections 1 and 2 of Chapter II, there); the strong, weak and $\sigma$-strong topologies were already introduced in [vN 2], although von Neumann referred to the $\sigma$-strong topology as the "strongest topology".

The double commutant theorem was proved in [vN 1], although the proof presented here is as in [Arv 1].

## Chapter 1

The discussion here is essentially a reproduction (although, probably, somewhat more cryptic) of Part II of [MvN 1]. The reduction theory that was briefly discussed at the end of Section 1.3 has its "fountainhead" in [vN 4].

## Chapter 2

The positivity of a linear functional on a $C^*$-algebra, which attains its norm at the identity element [cf. the parenthetical remark in Ex. 2.1.1 (c)] is proved in the Corollary to Theorem 1.7.1 in [Arv 1]. The equivalence, for positive linear functionals on a von Neumann algebra, of normality and $\sigma$-weak continuity (cf. the discussion following Ex. 2.1.4) is established in Theorem 1 in Section I.4.2 of [Dix]. Gleason's theorem (on the possibility of extending "measures" on $\mathcal{P}(\mathfrak{L}(\mathcal{H}))$ to normal positive linear functionals) appeared in: A. M. Gleason, Measures on closed subspaces of a Hilbert space, J. Math. Mech., **6** (1957), 885-894.

The fact that an injective *-homomorphism of $C^*$-algebras is necessarily isometric, as well as the consequent fact that a (not necessarily injective) *-homomorphic image of a $C^*$-algebra is norm-closed (cf. the discussion following Theorem 2.2.1) may be found in [Arv 1] (Theorem 1.3.2 and the discussion leading up to it). For a detailed discussion of cyclic sets and separating sets, see Section I.1.4 of [Dix].

The entire treatment of Section 2.3 is fashioned after that of Section 2.5.2 of [BR I]. The fact that $(M \otimes N)' = (M' \otimes N')$ (cf. the remarks in Example 2.3.7 (b)) is generally referred to as the "commutation theorem for tensor-products". It was first established by Tomita in: Quasi-standard von Neumann algebras, Mimeographed Notes, Kyushu Univ., 1967; also, it may be found as Theorem 12.3 in [Tak 1]; for a direct proof that does not appeal to generalized Hilbert algebras, the reader may consult [Sak].

Weights were introduced by Combes in [Com 1]. The method employed here (in Exercises 2.4.12 - 2.4.14) to establish that $\eta_\phi(N_\phi \cap N_\phi^*)$ has the structure of a generalized Hilbert algebra, is essentially the one to be found in [Com 2]. The entire discussion of generalized Hilbert algebras (in Section 2.4) is but a poor abstract of Sections 1-10 of [Tak 1], which the reader should consult for complete proofs (as well as for general self-improvement).

The discussion in Section 2.5 up to (and inclusive of) Corollary 2.5.12, is a modification of Section 13 of [Tak 1], where, however, only finite weights are considered. The K.M.S. characterization of the modular group of a semifinite weight (cf. Theorem 2.5.11') is due to Combes; see [Com 2] for a proof. All the material in Section 2.5 after Theorem 2.5.11' has its origins in [PT].

The treatment of Tomiyama's theorem (Prop. 2.6.4) is, but for language, exactly as in [Tak 4] (Theorem III.3.4). The complete proof of Theorem 2.6.3 (the Radon-Nikodym theorem for semifinite weights) may be found in [PT], where it first appeared. The version of Prop. 2.6.4 for semifinite weights (cf. Remark 2.6.9 (a)) is the main theorem of [Tak 2].

**Chapter 3**

The unitary cocycle theorem (Theorem 3.1.1 here) is Theorem 1.2.1 in [Con]; the proof given here is a reproduction of the one found there (although, probably, less cryptic in places). Theorem 3.1.6 of this book is from Section 14 of [Tak 1]; the proof presented here is an adaptation of that proof, taking into consideration the simplifying boundedness assumption made for the purposes of this proof.

Theorem 3.1.10 of this book is Theorem 4.6 of [PT].

The material on locally compact abelian groups (found here in the closed interval [Ex. (3.2.3), Prop. 3.2.6], as well as in the Appendix) may be found in any standard text on abstract harmonic analysis, such as [Rud] or [Loo] for instance. Barring that segment, all the material in Section 3.2 up to and including Lemma 3.2.8 is from [Arv 2]. The rest of Section 3.2 as well as all of Section 3.3 is from Sections 2.1 - 2.3 of [Con].

Almost all of Section 3.4 is from [Con] (Sections 3.1 and 3.2); the exceptions are: (i) Lemma 3.4 here is Prop. 2.3.17 in [Con]; (ii) the converse (Lemma 3.4.2) to the unitary cocycle theorem is from [Kal 2]; (iii) Sakai's result, used in the proof of Prop. 3.4.6, is Theorem 4.1.5 of [Sak]; and (iv) although [Con] contains Prop. 3.4.7, the explicit statement of Corollary 3.4.8 is not to be found there.

## Chapter 4

The crossed product of $M = L^{\infty}(X,\mathcal{F},\mu)$ by a countable group of automorphisms of $(X,\mathcal{F},\mu)$ was first considered in [MvN 1] and [vN 3], where versions of Corollary 4.1.9 and Prop. 4.1.15 can be found. The notion of an automorphism or an action being free (cf. Definition 4.1.8) as well as Prop. 4.1.11 originate from [Kal 2].

Krieger introduced the concept of a ratio set in [Kri], where he attributes the inspiration to Araki and Woods' notion of an asymptotic ratio set associated with what they term an ITPFI (cf. [AW]). The equality $S(L^{\infty}(X,\mathcal{F},\mu) \otimes_{\alpha} G) = r(G)$, as well as the conclusion of Ex. (4.2.13) were established in [Con], although the proof presented there is quite different from the one here.

The description (in Section 4.3) of factors of type $I_n$ $(1 \leqslant n \leqslant \infty)$ and $II_{\infty}$ (in terms of $II_1$ and $I_{\infty}$) may be found in [MvN 1] and [MvN 3], respectively; further, Theorem 4.3.13 and Examples 4.3.14 and 4.3.15 had their origins in [vN 3]. However, the presentation of the segment [Lemma 4.3.12, Example 4.3.15] is essentially as in [Tak 4]. The fact about $M_e$ being a factor when $M$ is, (used in the proof of Prop. 4.3.4 (b)), is the Corollary following Prop. 2 in Section I.2.1 of [Dix]. Corollary 4.3.19 is explicitly stated, at least when $G$ is cyclic, in [Arn].

All the material in Sections 4.4 and 4.5 first appeared in [Tak 3], except for the structure theorem for factors of type $III_{\lambda}$, $0 < \lambda < 1$ (stated as (2) at the end of Section 4.5), which came in [Con]. The treatment in Section 4.4 is, however, influenced by [vD]. Dixmier's result on the structure of isomorphisms of von Neumann algebras -- which was needed in the proof of Lemma 4.4.7 -- is in Section I.4.4 of [Dix] (particularly, see the Corollary and the Concluding Remark).

# BIBLIOGRAPHY


[AC] L. Accardi and C. Cecchini, Conditional expectations in von Neumann algebras and a theorem of Takesaki, J. of Functional Analysis, **45** (1982), 245-273.

[Arn] L. K. Arnold, On $\sigma$-finite invariant measures, Z. Wahrsch. verw. Geb., **9** (1968), 85-97.

[Arv 1] W. B. Arveson, An invitation to $C^*$-algebras, Springer, New York, 1976.

[Arv 2] W. B. Arveson, On groups of automorphisms of operator algebras, J. of Functional Analysis, **15** (1974), 217-243.

[AW] H. Araki and E. J. Woods, A classification of factors, Publ. Res. Inst. Math. Sci. Kyoto, **4** (1968), 51-130.

[BRI] O. Bratteli and D. W. Robinson, Operator algebras and quantum statistical mechanics I, Springer, New York, 1979.

[Com 1] F. Combes, Poids sur une $C^*$-algebra, J. Math. Pures et Appl., **47** (1968), 57-100.

[Com 2] F. Combes, Poids associes a une algebre hilbertienne a gauche, Comp. Math., **23** (1971), 49-77.

[Con] A. Connes, Une classification des facteurs de type III, Ann. Scient. Ecole Norm. Sup., **6** (1973), 133-252.

[vD] A. van Daele, Continuous crossed products and type III von Neumann algebras, Cambridge University Press, Cambridge, 1978.

[Dix] J. Dixmier, von Neumann algebras, North-Holland, New York, 1981.

[Haa] U. Haagerup, Normal weights on $W^*$-algebras, J. of Functional Analysis, **19** (1975), 302-318.

[Hal 1] P. R. Halmos, Measure Theory, Springer, New York, 1974.

[Hal 2] P. R. Halmos, A Hilbert space problem book, Springer, New York, 1967.

[Kal 1] R. R. Kallman, Groups of inner automorphisms of von Neumann algebras, J. Functional Analysis, **7** (1971), 43-60.

[Kal 2] R. R. Kallman, A generalisation of free action, Duke Math. J., **36** (1969), 781-789.

[Kri] W. Krieger, On the Araki-Woods asymptotic ratio set and non-singular transformations of a measure space, Lect. Notes in Math. 160, Springer, New York, 1970, pp. 158-177.

[Loo] L. H. Loomis, An introduction to abstract harmonic analysis, Van Nostrand, Princeton, 1953.

[MvN 1] F. J. Murray and J. von Neumann, On rings of operators, Ann. Math., **37** (1936), 116-229.

[MvN 2] F. J. Murray and J. von Neumann, Rings of operators II, Trans. Amer. Math. Soc., **41** (1937), 208-248.

[MvN 3] F. J. Murray and J. von Neumann, Rings of operators IV, Ann. Math., **44** (1943), 716-808.

[vN 1] J. von Neumann, Zur Algebra der Funktional operationen und theorie der normalen Operatoren, Math. Ann., **102** (1929), 370-427.

[vN 2] J. von Neumann, On a certain topology for rings of operators, Ann. Math., **37** (1936), 111-115.

[vN 3] J. von Neumann, On rings of operators III, Ann. Math., **41** (1940), 94-161.

[vN 4] J. von Neumann, On rings of operators. Reduction theory, Ann. Math., **50** (1949), 401-485.

[PT] G. K. Pedersen and M. Takesaki, The Radon-Nikodym theorem for von Neumann algebras, Acta Math., **130** (1973), 53-88.

[RS] M. Reed and B. Simon, Functional Analysis, Academic Press, New York, 1972.

[Rud] W. Rudin, Fourier Analysis on groups, Interscience Publ., New York, 1962.

[Sak] S. Sakai, $C^*$-algebras and $W^*$-algebras, Springer, New York, 1971.

[Tak 1] M. Takesaki, Tomita's theory of modular Hilbert algebras and its applications, Springer, Berlin, 1970.

[Tak 2] M. Takesaki, Conditional expectations in von Neumann algebras, J. of Functional Analysis, **9** (1972), 306-321.

[Tak 3] M. Takesaki, Duality in crossed products and the structure of von Neumann algebras of type III, Acta Math., **131** (1973), 249-310.

[Tak 4] M. Takesaki, Theory of operator algebras I, Springer, New York, 1979.

[Tit] E. C. Titchmarsh, The theory of functions, Oxford University Press, Oxford, 1939.

[Yos] K. Yosida, Functional analysis, Springer, Berlin-New York, 1968.

# INDEX

The accompanying numbers generally refer to the page containing the first occurrence of the term.